# The Future of Food: Exploring Edible Plants for a Sustainable Diet

Nikita Patel • Hemant Bagul • R. Krishnamurthy

# The Future of Food: Exploring Edible Plants for a Sustainable Diet

Springer

Nikita Patel
Kishorbhai Institute of Agriculture Sciences
and Research Centre
Uka Tarsadia University
Bardoli, Gujarat, India

Hemant Bagul
Polytechnic in Agriculture
Navsari Agricultural University
Waghai (Dangs), Gujarat, India

R. Krishnamurthy
Kishorbhai Institute of Agriculture Sciences
and Research Centre
Uka Tarsadia University
Bardoli, Gujarat, India

ISBN 978-3-032-15026-4 ISBN 978-3-032-15027-1 (eBook)
https://doi.org/10.1007/978-3-032-15027-1



This Springer imprint is published by the registered company Springer Nature Switzerland AG
The registered company address is: Gewerbestrasse 11, 6330 Cham, Switzerland

*In the quiet rhythm of the earth's cycles, where roots delve deep and harvests rise, this dedication extends to the dreamers who plant seeds of change, the innovators who reimagine abundance, and the children who will one day harvest a world of plenty. To the one who holds my heart, whose love and care fuels every chapter of my life—may this book reflect the warmth we've shared. For my family, whose meals taught me the joy of sustenance; my mentor, who guided me with wisdom and vision; and every farmer, chef, scientist, researcher and student pushing boundaries—thank you for the inspiration that filled these pages.*
*To the global community of eaters, growers and thinkers—past, present and future—who believe in a food system that heals rather than harms. Together, we can cultivate a legacy of nourishment for generations, where plates are full, choices wise and our planet thriving.*
*With eternal gratitude to God and those who paved the way.*

# Foreword I

In an age where the imperatives of ecological stewardship intersect with the imperatives of human nourishment, *The Future of Food: Exploring Edible Plants for a Sustainable Diet* emerges as a seminal exploration of our botanical heritage. I am delighted to endorse this work. Their erudite analysis transcends mere discourse, offering a profound synthesis of historical wisdom, scientific inquiry, and practical application.

The volume elucidates the transformative potential of edible plants, from the resilient wild greens that sustained ancestral civilizations to the cutting-edge biotechnologies reshaping modern cultivation. It articulates how these species not only fortify nutritional profiles and mitigate chronic ailments but also enhance soil vitality and climatic adaptability. Through meticulous case studies—ranging from indigenous crop revitalization to urban permaculture initiatives—the authors demonstrate the symbiotic relationship between diverse plant diets and global culinary traditions, fostering both individual well-being and communal resilience.

Moreover, the book serves as a catalyst for advocacy, equipping educators, policymakers, and practitioners with the tools to propagate sustainable practices. Its emphasis on education underscores the ethical imperative to cultivate knowledge alongside crops, ensuring that future generations inherit a thriving biosphere.

This is not merely a treatise; it is a clarion call for paradigm shift. I commend the authors for their scholarly rigor and visionary outlook. May this foreword inspire readers to engage deeply with the text and, in turn, with the edible wonders of our world.

Professor, Cape Peninsula University of Technology
Cape Town, South Africa

Learnmore Kambizi

# Foreword II

In a world grappling with the twin crisis of climate change and food insecurity, the resurgence of interest in edible plants offers a beacon of hope and a path to renewal. As a lifelong advocate for sustainable agriculture and a researcher who has witnessed the devastating impacts of monoculture farming on our ecosystems, I am thrilled to introduce *The Future of Food: Exploring Edible Plants for a Sustainable Diet*. This timely volume, penned by Dr. Nikita Patel, Dr. Hemant Bagul and Prof. Dr. R. Krishnamurthy, arrives at a pivotal moment when our food systems demand radical rethinking.

What sets this book apart is its holistic embrace of edible plants—not merely as ingredients, but as living threads connecting human health, cultural heritage, and planetary resilience. From the forgotten wild greens of ancient diets to the biotechnological marvels of modern vertical farming, the authors weave a compelling narrative that challenges us to look beyond the supermarket aisle. They remind us that edible plants are not relics of a bygone era but dynamic allies in combating chronic diseases, restoring soil health and fostering climate adaptation. Through vivid examples of urban gardening movements and global culinary traditions, the book illustrates how diverse plant-based diets can transform communities, one seed at a time.

Drawing on rigorous research and real-world case studies, *The Future of Food* bridges the gap between science and everyday action. It empowers readers—whether home gardeners, policymakers or curious eaters—to envision and enact a more sustainable future. The authors' passion for education and advocacy shines through, urging us to reclaim our relationship with food as a force for good.

This is more than a book; it's a manifesto for change. I compliment the book authors for their insightful work and encourage you to dive in. May it inspire you to cultivate not just plants, but a healthier and more equitable world.

Director of Research and Dean PG Studies
Navsari Agricultural University
Navsari, Gujarat, India

T. R. Ahlawat

# Foreword III

In an era defined by unprecedented environmental challenges and a growing awareness of the fragility of our food systems, *The Future of Food: Exploring Edible Plants for a Sustainable Diet* arrives as a timely and indispensable guide. As a leading advocate for sustainable agriculture and global food security, I am honored to introduce this groundbreaking work, which offers a compelling vision for reimagining our relationship with the natural world through the lens of edible plants.

The book masterfully traces the "edible plant revolution," reminding us that these humble yet powerful organisms have sustained humanity for millennia. From ancient grains and wild greens to indigenous crops, the authors illuminate how rediscovering underutilized species can nourish both body and planet. They delve into the nutritional prowess of plant-based diets, demonstrating their role in combating chronic diseases while fostering sustainability. Innovations like precision farming, vertical agriculture, and lab-grown foods are explored with clarity, revealing how technology can harmonize with tradition to build resilient food systems.

What sets this volume apart is its holistic approach: it bridges science, culture, and community. Chapters on climate resilience showcase real-world examples of diverse crops adapting to environmental shifts, while the urban gardening movement empowers readers with practical tools for local action. The authors' global perspective celebrates culinary traditions, from Mediterranean wild herbs to Asian vegetables, underscoring how plant-based diets have long promoted health and ecological balance.

The researchers concluded that a healthy plant-based diet might reduce the burden of multimorbidity of cancer and cardiometabolic diseases. The study also reported evidence of how plant-based diets is also environmentally more sustainable than diets with a higher preparation of animal products. The researchers were able to associate a higher adherence to a healthy plant-based diet with a lower rich of cancer and cardiovascular diseases. India's scientists should generate ideas to move from food security to nutrition security, create biofortified crops to address malnutrition, develop low-cost fertilizers better map India's genomic biodiversity for personalized medicine, and arrive at new and cheap innovation in clean battery storage.

Yet, the book is not merely academic; it is a call to action. Through education and advocacy, it inspires schools, workplaces, and communities to embrace sustainable eating. As someone who has witnessed the transformative power of plant diversity in initiatives like community seed banks and agroforestry projects, I can attest that this knowledge is not just informative—it's revolutionary.

This book is essential reading for students, researchers, educators and anyone committed to a healthier, more equitable world. May it inspire a global shift toward diets rooted in abundance and respect for the Earth.

Professor and Head (Retd.,) A. Panneerselvam
A V. V. M Sri Pushpam College (Autonomous)
Thanjavur, Tamil Nadu, India
Honorary Visiting Expert in Dept. of Biology, GRI (DTBU)
Dindigul, Tamil Nadu, India

# Preface

In an era defined by rapid environmental change, shifting dietary habits, and the urgent call for sustainable living, the humble plant stands as both our ancestral ally and our greatest hope for the future. As we embarked on this journey to write *The Future of Food: Exploring Edible Plants for a Sustainable Diet*, we were struck by the stark contrast between the abundant wealth of nature's edible plants and the industrialized food systems that dominate our plates. From the wild greens foraged by our ancestors to the cutting-edge innovations of vertical farming/skyfarming, edible plants hold the key to nourishing our bodies, healing our planet and preserving our cultural heritage.

This book is born from a deep conviction that food is not merely sustenance but a powerful force for transformation. Drawing on historical insights, scientific advancements, and global perspectives, it invites readers to rediscover the vast world of edible plants—those familiar staples and the lesser-known treasures that have sustained humanity for millennia. Through its chapters, we explore the resurgence of interest in these plants, their nutritional prowess, and their role in building resilient, equitable food systems. From the health benefits of diverse plant-based diets to the technological marvels reshaping agriculture, and from urban gardening initiatives to cultural traditions around the world, *The Future of Food* serves as a guide for anyone seeking to eat more mindfully and live more sustainably.

Whether you're a home gardener, a policy advocate, or simply curious about the flavors and futures of plant-based eating, this book aims to inspire action. It is our hope that by the end, you'll see edible plants not as relics of the past, but as vibrant threads weaving a healthier, more harmonious world. Thank you for joining us on this exploration—may it cultivate change, one seed at a time.

Bardoli, Gujarat, India — Nikita Patel
Waghai (Dangs), Gujarat, India — Hemant Bagul
Bardoli, Gujarat, India — R. Krishnamurthy

# Contents

# About the Authors

**Nikita Patel** is a Research Associate at the Kishorbhai Institute of Agriculture Sciences and Research Centre, Uka Tarsadia University, Bardoli, Gujarat, India. She completed her PhD at the C.G. Bhakta Institute of Biotechnology (CGBIBT) at the same University, where she also earned her Bachelor's and Master's degree in Biotechnology with first-class honors.

With experience as a lecturer at both the Dollyben Desai Institute of Computer and Allied Science and CGBIBT, Dr. Nikita has actively contributed to the academic community. She is also working as an Instructor at the Department of Health Science, University of the People, Pasadena, USA. She has presented her research at various national and international conferences and has attended several workshops and training programs. She is also serving as an editorial and review board member of many national and international journals. Her research findings have been published in international journals, including those by Springer, Elsevier and Taylor and Francis. Additionally, she has edited a book titled *Controlled Environment Horticulture: Where Plant Biotechnology Meets Sustainable Future* by Bentham Science and holds two Indian design patents.

In recognition of her contributions to the field, Dr. Nikita was conferred with the Young Scientist Award in 2022. Her primary research interests include Agricultural Biotechnology, Microbial Biotechnology, Plant Biotechnology and Molecular Biology, highlighting her commitment to advancing knowledge in these vital fields.

**Hemant Bagul** is currently serving as an Assistant Professor at the Polytechnic in Agriculture, Navsari Agricultural University, Waghai (Dangs), Gujarat. He holds a B. Sc. (Hons.) in Horticulture, an M. Sc., and a Ph.D. in Fruit Science from the ASPEE College of Horticulture and Forestry, NAU, Navsari, Gujarat, and has qualified the ICAR National Eligibility Test (NET). During his PhD, he received a National Fellowship and Scholarship for Higher Education for his research in Fruit Science.

In recognition of his contributions, Dr. Hemant was conferred with the Young Horticulturist Award in 2024. He actively serves as an editorial and review board member for numerous national and international journals and has authored several book chapters, research papers, and articles published in reputed journals. He has also edited a book titled *Controlled Environment Horticulture: Where Plant Biotechnology Meets Sustainable Future* by Bentham Science. Beyond his academic pursuits, Dr. Hemant is an accomplished athlete, having represented as an eight-time national player in kho kho and athletics (running 1500 m). His primary research interests include fruit science, crop production and tribal agriculture, reflecting his dedication to advancing agricultural knowledge and practices.

**R. Krishnamurthy** is Professor, Research Director, Uka Tarsadia University; Director, Kishorbhai Institute of Agriculture Sciences and Research Centre; and Dean of Faculty of Science, Uka Tarsadia University (UTU), Bardoli, Gujarat, India. He was conferred Man of Excellence by Indian Achievers' Forum in 2022 and Fellow Status Award by Chartered Institute of Environmental and Public Health Management of Nigeria (CIEPHM) in April 2017 at Lagos. He is a great visionary, research mentor, and a very meticulous administrator. He impeccably executes multidimensional responsibilities viz. teaching, research, and associated administrative activities with enthusiasm. He is a Visiting Professor to the University of Ilorin (Unilorin), Nigeria, and an Associate Research Mentor for Nigerian students in their academic research work since March 2013. He has facilitated over 18 academic staff of Unilorin from various departments to visit Uka Tarsadia University with granted research fellowships, including

NG Patel Research Fellowship (from Lagos), and collaborated as mentor to complete their project. From 2023 to 2024), three academic faculties from Unilorin are conducting research and academic activities under his mentorship. He has published over 35 research articles along with Unilorin academic staff and published 8 book chapters published by leading international publishers. He has also guided a Unilorin staff to receive PhD from UTU and other four staffs to complete their PhD work at UTU and submitted their thesis to Unilorin. In addition to these, Prof. Krishnamurthy has facilitated and arranged inter universities faculty lecture earlier where vice chancellor of Unilorin and various departments' professors and HODs including former CREDIT and CIE directors had lectured in the event. He has completed his Master degree in Botany from the University of Madras and PhD (Botany), MS University, Baroda. He is a recipient of several research fellowships, awards, and grants. He served as member of academic bodies in universities in India. He has a vast research experience on medicinal plants for decade with industry as director of research and over 18 years with academic institutions. He has technologically helped Indian and overseas industries and universities including Unilorin to set-up commercial scale plant tissue culture laboratories for quality planting materials of plants. Dr. Krishnamurthy helped Unilorin to cultivate and develop upland rice farming at Unilorin campus and same project is still in progress. He has organized (as chairman) several national and international seminars/conference on medicinal plants and food security plants. He has to his credit a European patent and written and edited several book chapters on medicinal plants and more than 140 research publications and review articles in the areas of medicinal and food plants. He is a reviewer of many international journals and member of scientific societies. He collaborates with overseas universities for research work. Expanding his horizon and not keeping himself limited to the academia, he ventures in helping farmers in cultivating medicinal plants.

# Chapter 1
# The Edible Plant Revolution

**Abstract** The Edible plant revolution identifies and discusses the renewed global interest in edible plants and their redemptive power to transform diet and food systems. This chapter explores how edible plants which were once staple of human nutrition, are making comeback as a sustainable and culturally important food source. It explores the history of edible plants starting from human diet throughout civilizations and have constructed cultural identity through traditional practices. By revisiting the significance of edible plants, the chapter underscores the capacity for responding to contemporary challenges like food security, health and environmental sustainability. It also highlights the importance of integrating edible plants for transforming modern food system by bridging people, culture and nature more profoundly.

**Keywords** Case Studies · Edible Plants · History · Modern Diet · Revolution · Traditional Diet

## 1.1 Introduction

In recent years there has been a remarkable resurgence of interest in edible plants driven by growing awareness of health, sustainability and food security. Edible plants encompass a wide array of species including fruits, vegetables, grains, legumes, nuts, seeds and even lesser known wild plants such as foraged greens and ancient grains. This renewed focus highlights their potential to transform diets and food system worldwide (Rasingam 2012). The edible plants offer numerous benefits as they are rich sources of essential nutrients, fiber, vitamins minerals and phytochemicals that promote health benefits and prevent chronic diseases such as diabetes, cancer and heart diseases. For example, leafy vegetable or leafy greens like kale and spinach are high in antioxidants and vitamin A, C and K while legumes such as lentils and chickpeas provide plant based protein and fiber (Barwant 2021).

Moreover, plant based diets tends to have a lower environmental footprint compared to animal based diets. The production of one kilogram of beef approximately



N. Patel et al., *The Future of Food: Exploring Edible Plants for a Sustainable Diet*,
https://doi.org/10.1007/978-3-032-15027-1_1

requires 15,000 liters of water whereas many plant food require significantly less. Additionally, agriculture generally emits fewer greenhouse gases and uses less land, making edible plants a cornerstone of sustainable food systems (Willett et al. 2019).

In addition to serving as food, plants are essential to our civilizations, religions, and medical practices. One example of this is how the exquisite curvature of a tendril can serve as an inspiration for art, or how native forest dwellers gather plant resources for religious or therapeutic purposes. Plants are essential to our society and provide us with more than simply sustenance (Raven 2019). India is the second largest country in the world by populace. Approximately 75% of Indians reside in rural areas, where they rely only on wild plant sources to supplement their diet and meet their nutritional demands (Rashid et al. 2008). An estimated 800 distinct varieties of food plants are consumed in India. Humans have been using wild plants as food since the dawn of time. They are used for many different purposes, some of which are used as food. While some wild plants are used as domestic animal feed, others are used as fibre and medicine. Humans use wild plants as food because of their nutritional value. Emerging nations are hence dependent on agricultural output. In rural communities, especially during periods of food scarcity, indigenous food plants and certain agricultural plants serve a variety of purposes, including the production of medicine and as supplements to reduce hunger (Vishwakarma and Dubey 2011).

## 1.2 Plants Through the Ages: A Historical Perspective

We often forget that wild plant diets were crucial to the evolution of primates, including humans, in our agriculturally oriented global economy. Additionally, some of the few remaining hunter-gatherer societies are still supported by wild resources. From the emergence of the first humans millions of years ago, hunter-gatherers have lived in a wide range of climates and environments, developing survival skills via the use of an astounding assortment of plants and animals (Bocquet-Appel 2009).

The evolution of primate species was fueled by the proliferation of flowering plants. The Angiosperms, or flowering plants, had established themselves by 65 million years ago, towards the end of the Cretaceous period, and broad-leafed, fruit-bearing trees started to take over the enormous forests that eventually covered a large portion of the planet (Torija-Isasa and Matallana-González 2016).

Fossil fruits and seeds suggest that species akin to the modern sweet sop (*Annona squamosa*) and sour sop (*Annona muricata*) dominated the inland woods, with mangrove and swamp palms predominating in coastal areas. Mango, walnut, and pistachio early varieties seem to have existed. Palms, Sequoia conifers and climbing plants like vines and lianas coexisted with trees like bay, cinnamon, magnolias, and black gum trees. The nectar and nourishment from the blooming plants had already helped the birds adjust to this transition. A larger variety and quantity of food were made available by the new plants (Rivera et al. 2006). As a result of eating the fruit, the birds assisted in spreading the seeds via their feet, beak, and excrement, which was a win-win situation. By spreading pollen from plant to plant in their quest for nectar, insects already actively participated in this interaction. The fact that no single plant can supply all of the amino acids the body needs as building blocks to generate animal protein is a significant drawback of consuming only plant-based diets. To guarantee that the diet has all of the required amino acids, a wide range of plants must be consumed. This is demonstrated by the way chimpanzees behave, as their dietary requirements frequently force them to hunt opportunistically (Leonti et al. 2006).

This is due to the fact that, in contrast to plant products, animal meat includes every amino acid required by any other organism to synthesize its own body tissue. Because of this, chimpanzees' diet is even more varied than that of humans, which significantly reduces the range of environments in which they may live. As a matter of fact, it is among the primary causes of their extreme extinction. However, being scavengers and perhaps opportunistic hunters of small animals while living on the edges of the African savanna is likely what allowed our own ancestors, the slim australopithecines, to survive after the deaths of their strong counterparts (McGrew 1983).

## 1.3 Indigenous Epistemologies: System of Knowledge and Practice

The return of Indigenous land and life is the result of decolonization; it is not a metaphor for other things we wish to do to improve our societies and schools. It is necessary for generations of Indigenous Peoples to be deeply and firmly rooted in their homelands, encased in their languages and religious beliefs, and embodying their traditions of agency, leadership, decision-making, and diplomacy in order for Indigenous political cultures, governances, and nation-building to flourish. People all throughout the world have viewpoints on how humans interact with other living forms and ecosystems, as well as knowledge that biologists can study and that is extremely valuable (Tuck and Yang 2012). However, people who have lived in a particular area for many generations—sometimes even for millennia—have unique knowledge about the species that call that area home, the changes that have occurred over time, and the intimate, mutually dependent relationships that exist between humans and other living forms. Indeed, these folks are "well grounded." Because of this, Indigenous peoples' and other long-resident peoples' viewpoints and expertise are extremely pertinent to ethnobiological/biological research (Turner et al. 2012).

Over the past few generations, foreign impacts have caused significant dietary changes for many Indigenous Peoples worldwide, leading to widespread instances of weakened overall health, including an increase in type II diabetes, obesity, heart problems, and high blood pressure, as well as a correspondingly lower life expectancy (Joseph 2021). Ethnobiological research on Indigenous food systems, traditional healing methods, and the general well-being of Indigenous populations, based on cultural knowledge and related studies, has made significant contributions to our understanding of these issues and potential solutions, in addition to the revival of traditional knowledge and the reclamation of land-based rights and practices occurring within Indigenous communities. This covers research on agricultural cultivars, genetic selection, and indigenous and local food biodiversity (Joseph and Turner 2020).

Indigenous Peoples have faced numerous obstacles in accessing and utilizing their foods and engaging in their cultural practices due to the activities of governments, industries and laws and rules implemented by European newcomers for their personal profit (Kimmerer 2013). These attempts have been made worse by unforeseen and unanticipated effects imposed by a variety of other shocks brought on by the arrival of the foreigners, including invasive species, diseases, insect pests, pollinator impacts, deforestation, biodiversity loss, contamination and destruction of wetlands, and, most recently, climate change (Geniusz 2015).

For those who are affected, the loss of food and the inability to harvest it has caused a great deal of sorrow and distress. In her book "What Do Clams Taste Like?," a Gitka'a'ata grandmother wrote about her granddaughter who had never tried abalone because it was protected from harvesting. She also wrote about children's anxieties about the future and how they would only ever be able to hear stories about their Indigenous foods rather than taste them.

Foods from the past have memories and are frequently closely associated with identity. Reminiscences of time spent on the land with loved ones can be closely linked to specific flavours, scents and textures. Connections to places, people, culture and spirituality are facilitated by ancestral foods (Satterfield et al. 2012). Despite all of these adverse developments and effects, Indigenous communities have shown resiliency and fortitude. This resilience is demonstrated by the resurgence of ancestral food-related cultural knowledge and practices occurring in many Indigenous communities (Talaga 2018). Restoring management and sustainability methods related to ancestral foods is an everyday act of Indigenous resurgence as well as a political act. Indigenous cultural and political resurgence today is a tribute to the generations that battled to preserve aspects of their culture, language, identity, and integrity so that future generations could progress toward improved health, a stronger bond with the land, and a more vibrant sense of cultural identity, both within the study area and beyond (Elliott 2019).

## 1.4 Domestication and Co-Adaptation

One of the most significant developments in human history is the domestication of plants. Hunter or gatherers relied on wild plants more than 20,000 years ago because there were no cultivated plants. These wild species became domesticated or genetically changed, during their time with humans, and ultimately evolved into crop species that differ greatly from their wild counterparts (Fig. 1.1). Rice cultivation

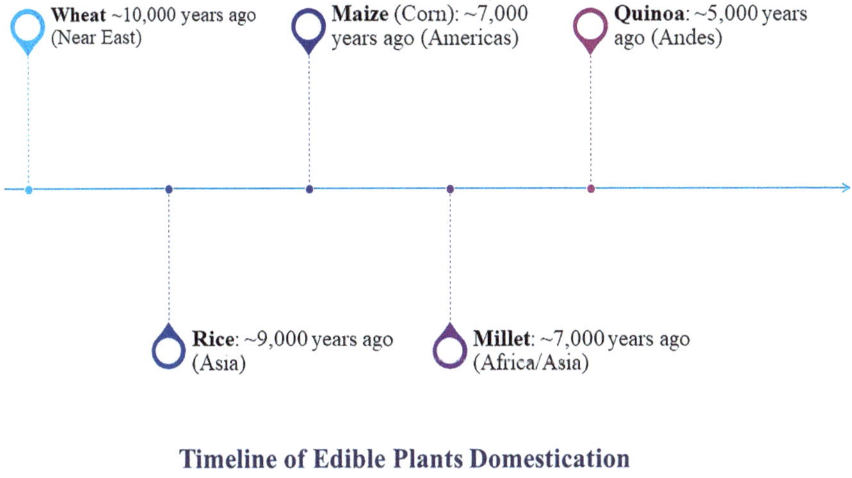


**Fig. 1.1** Timeline of Edible Plants Domestication

has a wild predecessor, called *Oryza rufipogon*, but these two plants are quite distinct (Sweeney and McCouch 2007). *Oryza rufipogon* generates long awns to help disperse seeds and very big anthers for cross pollination, both of which are absent from the present crop. These significant changes in appearance are linked to domestication because the seeds of the majority of farmed rice varieties are fertilized before the blossom opens. Many animals are rendered incapable of surviving and competing in the wild due to the genetic alterations brought about by domestication (Kovach et al. 2007; Fornasiero et al. 2022). They are now reliant on people as a result. Throughout history, domestication has taken place on numerous occasions. Multiple centres of domestication exist; for instance, domesticated rice and soy beans (*Glycine max*) are found in Asia. Yam (*Dioscorea* sp.) is an example of domestication in Africa, whereas wheat and barley (*Hordeum vulgare*) were domesticated throughout the Fertile Crescent, and maize and potatoes (*Solanum tuberosum*) were farmed in the Americas. The native peoples of America domesticated a wide variety of species in the eastern United States (Grun 1990; Zohary 2017; Sedivy et al. 2017).

The way our forefathers used the land was altered by our capacity to domesticate and grow plants near to where we now reside. They started establishing permanent communities, population growth, and the division of labor as food became more dependable. A cultural transition resulted from the fact that not all people produced food and not everyone was from hunter-gatherer community (Pickersgill 2007). Mesoamerica is the place where maize's wild ancestor, Teosinte (*Z. mays* ssp. parviglumis), was domesticated. Similar to *O. rufipogon* and rice, the fruiting structures of maize and teosinte differ significantly. Domestication again produced striking morphological changes because teosinte is a large perennial with numerous flowers and flowering branches, whereas the domesticated maize plant has a few large leaves, two ears and a single stalk. Human-influenced selection, also known as

Darwin's "artificial selection," is responsible for the distinctions between domesticated crops and the wild counterparts. The first cultivators noticed that certain plants had superior qualities than others, such as better-tasting, more seeds that were retained, or easier to cultivate or more vigorous. They also grew wild plant seeds near their homes. Until the plant's character was altered, the first cultivators would gather the seeds of the best plants to produce the following year. This domestication process continued year after year, generation after generation (Casas et al. 2007).

Agricultural domestication is linked to entire sets of traits. In domesticated plants, seed dispersal is frequently lost, especially when the seeds are the part we eat and must thus stay on the plant in order to be harvested. Dispersal mechanisms were swiftly eliminated during domestication because farmers do not want seeds to fall from the mother plant, even though wild plants use them to find and colonize new locations (Sangeetha et al. 2020). In Thailand, a number of rice cultivars exhibit intermediate domestication, and some seeds keep their long awns for propagation. For domesticated plants to be harvested at the same time, they must also blossom concurrently. Compared to domesticated types, wild plants typically have more stems, fewer seeds, and are taller (Castillo 2011).

Humans frequently adopted plants that were both safer and delicious; for instance, several chemicals generated from SA can be bitter and interfere with digestion; hence, these traits were removed during domestication. The genetic foundation of domestication is now well understood. Due to the activity of the Teosinte Glume Architecture (TGA1) gene, Teosinte has a very hard kernel; however, during the domestication process, a tga1 mutation was used to make maize's glumes softer and smaller, making the kernels much simpler to harvest and consume (Buckler and Stevens 2006). Teosinte yields numerous branches with multiple inflorescences, whereas in maize, a mutation in TB1 stops lateral branches from developing, leaving only one stalk. Instead of the enormous vegetative state seen in teosinte, this allows all of the plant's energy to be directed on producing the single ear of corn found in maize. Modern maize lacked the genetic diversity of Teosinte. Through a comparison of ancient DNA and some traditional varieties, the researchers were able to ascertain the path of maize domestication; the wild ancestor possessed numerous alleles at a single locus, allowing humans to select the most desirable traits (Xue et al. 2016).

Three of the nine necessary amino acids that are not synthesized in the human body are biosynthesised by plants via the shikimic acid pathway, which is perhaps one of the most significant metabolic pathways in plants. Humans now rely on plants because they have evolved which needs derivatives of this pathway. Additionally, this pathway produces flavonoids and alkaloids, many of which are utilized as medications. For instance, sweetgum (*Liquidambar styraciflua*) and star anise (*Illicium verum*) are sources of the pathway, which is the building block for Tamiflu, a medication used to prevent the severe symptoms of influenza. The intimate evolutionary relationship between plants and humans, which is based on biochemistry, is further demonstrated by these food and medicinal uses.

Without absorbing the essential amino acids from our diet, we are unable to biosynthesize all of the amino acids our bodies need to produce the entire proteins required for bodily functions. Cereals have the exact opposite composition to legumes, which are very high in some amino acids and low in others. A noteworthy nutritional trait that has frequently surfaced in various human civilizations with the ability to obtain a complete set of amino acids by mixing these two plant types. Ascorbate, also known as Vitamin C, is another intriguing case in this regard. Because humans cannot generate Vitamin C on their own, primitive societies relied on citrus fruits and other species to prevent ascorbate deficiency, which can lead to scurvy (Smirnoff 2011; Martin and Li 2017).

## 1.5 Case Studies

### *1.5.1 The Rediscovery and Agronomic Revival of Quinoa in Andean Region*

Quinoa (*Chenopodium quinoa* Willd.), an ancient food crop, is gaining popularity due to its excellent nutritional value, amazing yields, and high adaptation to various pedoclimatic situations. Quinoa is regarded as a pseudocereal and is a member of the Chenopodiacee family. Since 5000 years ago, it has been a native food plant in the Andes. Quinoa was domesticated between 3000 and 4000 years ago and originated in the Andean region of Bolivia, Colombia, Ecuador, and Peru (Jancurová et al. 2009; Tapia 2015).

For the Andean peoples, this crop was crucial for food and animal feed. It was also revered as the "mother grain" in their rituals. Spanish conquistador Francisco Pizarro and his troops forced Old World religion and customs, such as Catholicism, potatoes, wheat, and barley, on Inca villages after they overran the Andes in 1532. The Incas continued to produce quinoa in secret places despite the possibility that they would be slaughtered or have their hands severed for doing so (Jacobsen et al. 2003). Quinoa has reportedly been grown continually in a particular area for a longer period of time than any other crop (Andrews 2017).

Due to the "rediscovery" of the quinoa plant's nutritional benefits, agronomic traits, and exceptional tolerance to various pedoclimatic and water conditions, interest in the plant grew in the second half of the previous century. Quinoa's tale is remarkably similar to those of dozens of other crops that the Incas developed but that Spanish colonists ignored, such as amaranthus. Conquistadors preferred to massacre them in order to outlaw them since they were cheap plants, ignored their nutritional benefit, and were terrified of their ceremonial significance. These crops have long been reassessed and placed into a brand-new, sizable family known as the "Lost Crops of Incas" (McDonell 2020).

Despite the fact that the Spanish are remembered for their contribution to the spread of "new crops," such as maize, tomatoes, peanuts, potatoes, etc. Of the more than 200 types of quinoa, the real quinoa is the most widely used, primarily in the brackish regions of the Salar. Quinoa is an annual, robust herbaceous plant in the areas of Potosi and Oruro. Quinoa's great nutritional value was valued by the Incas and other ancient tribes, who included its leaves and seeds in their diet in order to make up for the lack of animal proteins. Quinoa seeds are high in protein and gluten-free, making them beneficial for celiac disease diets (Winkel et al. 2018).

The United Nations (ONU) has designated 2013 as the "International Year of Quinoa" to highlight the grain's importance as a global food security crop due to its nutritional, economic, environmental, and cultural value. Because of its health advantages and capacity to withstand challenging growing circumstances, quinoa has recently experienced a global renaissance after centuries of neglect during colonial times. This example shows how ancient edible plants can support contemporary nutrition and food security (González et al. 2020).

### *1.5.2 The Resurgence of Teff in Ethiopia*

Ethiopia is a culturally diverse country with around 80 distinct ethnic groupings. Teff is an important component of Ethiopia's national identity and has a long history dating back to Ethiopian antiquity. The ancient, self-pollinating warm-season annual grain teff is used to make the Ethiopian staple bread injera (Crymes 2015).

Teff was initially domesticated in Ethiopia between 4000 and 1000 BC, according to Nicolai Vavilov. Africans gathered the seed grains of local wild grasses for food and selected the traits that best fit their preferences, farming methods, and growing conditions, transforming teff into one of the grain crops that made the transition from wild grasses to domesticated food (Firew 2010). Teff grain is quite small and can range in color from light tan to dark brown to reddish-brown purple, as well as pale white to ivory white. Teff can be harvested 2–5 months after seeding, depending on the type. One kernel of wheat is equivalent to 150 grains of teff, making it the tiniest grain in the world. In Ethiopia, the primary teff-producing regions are Gojam and Shewa (in the central highlands), Gonder, Wello, and Welega. Although teff is typically grown in the Ethiopian highlands, it is currently being grown in a variety of environments, including marginal soils and floods. This adaptability may help to explain why teff is currently grown in a variety of locations, including the low wetlands of the Netherlands and the arid highlands of Idaho (Assefa and Chanyalew 2018).

The NRC claims that Wayne Carlson brought teff to the US for the first time in the 1980s, and that it is now being cultivated in Idaho for the enormous Ethiopian Diasporan communities. To reseed the arid and semi-arid rangelands of the southwest, however, numerous species of Eragrostis were first brought to the United States in the early 1930s, according to Streetman (1963).

Numerous sources, however, disagree over the exact date of Teff's worldwide activities. The Royal Botanic Gardens, Kew, imported seed from Ethiopia in 1866 and dispersed it to the United States, India, Australia, and South Africa, according to Ketema's (1997) monograph. In 1911, Skyes brought it to Zimbabwe, Mozambique, Kenya, Uganda, and Tanzania, and in 1940, Horuitz brought it to

Palestine. Burt Davy brought teff to California, Malawi, Zaire, India, Sri Lanka, Australia, New Zealand, and Argentina in 1916, according to Ebba (1975) (Dijkstra et al. 2008).

### *1.5.3 Rediscovery of Amaranth Cultivation Among Indigenous Communities*

Before the Spanish Conquistadors upended South American culture, 5000–7000 years ago, the Aztecs cultivated amaranthus species as their main grain crop. Terms like "super grain of the Aztecs," "mystical grains of the Aztecs," and "golden grain of the Gods" were used to refer to the nutrient-dense amaranth grain. It was seen that the grain gave newborns nourishment and gave troops vigor and energy throughout long journeys. It would take six centuries for contemporary biology to validate the early civilizations' experiential knowledge of these nutritional variables (Marx 1977).

Although amaranths were the primary grain utilized in the South American continent, early civilizations cultivated them as a vegetable crop more than 2000 years ago, and they are still used almost everywhere in the world today. A wide variety of geographical regions, including the southwestern United States, China, India, Africa, Nepal, South Pacific Islands, Caribbean, Greece, Italy, and Russia, have used and continue to use vegetable Amaranthus spp. for human consumption (Fomsgaard et al. 2010).

Although different species of grains and vegetables can be identified, it is common practice to use both the grain and the leaves of different sorts for animal and human use. Between 2000 and 3000 hectares are thought to be produced in America today, with the Great Plains region especially Nebraska producing the most, and the Midwest having several smaller production regions (Stallknecht and Schulz-Schaeffer 1993; Turner et al. 2021). Of the roughly 70 species of the genus Amaranthus, 40 are indigenous to the America, while the remaining species can be identified in Australia, Africa, Asia, and Europe. The amaranth species that have generated global interest as a crop in recent years are *A. caudatus*, *A. hypochondriacus* and *A. cruentus*. This could be as a result of its seeds and leaves having a high nutritious content (Sani et al. 2004).

The four areas in Mexico where amaranth is primarily produced for commercial use are Tulyehualco, Mixiquic, Tetelco and Tecómitl in the Federal District; Huaquechula, San Juan Amecac, Tochimilco, Tochimzolco, and Santa Clara Tetla in Puebla; San Miguel del Milagro in Tlaxcala; and Huazulco and Amilcingo in Morelos. Other states that occasionally engage in commercial production include Mexico, Oaxaca, Guerrero, Durango, and more recently, Chihuahua, Guanajuato, Jalisco, San Luis Potosí, and Michoacán (Espitia-Rangel et al. 2010).

There is still much to learn about the domestication of amaranth species for grain. The Coxcatlan cave archaeological excavations in Tehuacán, Puebla, Mexico, provide the earliest known information on amaranth as a domesticated grain crop. It was calculated that the seeds of *A. cruentus* were around 6000 years old. Although they were only thought to be 1500 years old, the earliest known *A. hypochondriacus* seeds were discovered in the same caves as *A. cruentus*. It's possible that *A. hypochondriacus* was domesticated earlier, most likely in the Valley of Mexico. Additionally, this species is cultivated in Southern India and Nepal's Himalayas, which have developed secondary diversification centres (Schnetzler 2018).

### 1.5.4 Role of Millets in Indian Agriculture

In India, millet crops are small-seeded grains cultivated for a long time. Millets are classified as minor cereals within the Poaceae family of grasses. Millets may grow in a variety of tropical and desert environments (Michaelraj and Shanmugam 2013). They can even survive in less fertile soil. In a resolution supported by over 70 countries and led by India, the United Nations (U.N.) has declared 2023 as International year of millet (Bandyopadhyay et al. 2017).

Ragi (finger millet), bajra (pearl millet), kangni (foxtail millet), and kutki (small millet) are only a few of the many possible variations. Millets are widely known for their capacity to adapt to a range of agroecological situations, which makes them ideal for growing in regions that experience both rain and drought. India may create a more robust and sustainable agricultural system that promotes food security, environmental preservation, and rural development by embracing millet crops (Dwivedi et al. 2023).

India is the world's leading millet producer. The two millets produced in India, Sorghum (Jowar) and Pearl millet (Bajra), would make up around 19% of the worldwide productivity in 2020. 40.51% of the world's pearl millet and 8.09% of its sorghum millet were produced in India in 2020 (Lenka et al. 2020). The top producing states in India for millets are Madhya Pradesh, Gujarat, Madhya Pradesh, Tamil Nadu, Andhra Pradesh, Maharashtra, Uttar Pradesh, Haryana, Rajasthan and Karnataka. Approximately 98% of India's millet production in 2020–21 comes from these ten states combined. More than 83% of the world's millet production is split among six states: Rajasthan, Karnataka, Maharashtra, Uttar Pradesh, Haryana and Gujarat. Rajasthan accounts for 28% of India's total millet production (Mishra 2023).

India faces numerous challenges in building sustainable agriculture because of its vast geographical region and diverse climate. The consequences of climate change, water scarcity, soil degradation, and the need to ensure food security for the nation's growing population are some of these challenges. In this case, cultivating millet crops is a viable solution to these issues and a way to promote agricultural sustainability (Lokesh et al. 2022).

Millets used to be an essential part of indigenous farming methods in India. They were valued for their ability to thrive in a variety of agroclimatic settings with minimal care or input. Traditionally, millets have been consumed in a number of forms, including fermented foods, whole grains, flour, and malted drinks. They have been an essential component of diets, improving the nutritional status of populations (Chaudhary et al. 2023).

Furthermore, millets have traditionally been utilized as a hedge against crop failure and erratic weather. Their ability to produce decent yields even in difficult conditions and their tolerance to drought have made them a reliable source of food security for small-scale farmers. Even though millet yield decreased as a result of the Green Revolution, there has recently been renewed interest in reintroducing millet crops. In India, there are efforts to promote their production and consumption because of their potential for sustainable agriculture, climatic tolerance, and nutritional value (Bagul et al. 2024). The importance of restoring and promoting millets for the nation's agricultural sustainability is highlighted by their historical significance in Indian agriculture, which provides a strong basis for acknowledging their potential as sustainable crops.

## 1.6 Conclusion

Living at the junction of environmental sustainability, cultural heritage and human health, edible plants are a template for resilient food systems in a time of worldwide crisis. Their value, as representatives of biodiversity, in feeding human populations while representing a protective status for ecosystems has been established since ancient whole food societies, and more recently as new technology opens up possibilities such as urban farming and climate-adaptive commodities. Various cases from American, Indian, African, and Andean land use demonstrates how the reincorporation of a broad array of edible species can alleviate the ecological burden of conventional agriculture and explore the theory of reduced food insecurity, malnutrition and global climate change. To fully realize the potential of edible plants, it will be crucial to prioritize the conservation of heirloom varieties, indigenous knowledge, and agroecological practices as we proceed forward. We can promote healthier societies, restore the planet back into balance, and respect the lasting legacy of our plant-based heritage by incorporating them into diets and policies around the world. In the end, adopting edible plants is a commitment to a sustainable future for future generations as well as a dietary decision.

## References

Andrews D (2017) Race, status, and biodiversity: the social climbing of quinoa. Culture Agric Food Environ 39(1):15–24

Assefa K, Chanyalew S (2018) Agronomics of tef. The economics of tef, exploring Ethiopia's biggest cash crop. International Food Policy Research Institute (IFPRI), Washington, DC, pp 39–70

Bagul H, Patel N, Patel J, Mangroliya RM, Bhoye VM (2024) The marvelous world of ragi: a nutritional odyssey through time and tradition. Acta Sci Agric 8(12):29–33. ISSN 2581-365X

Bandyopadhyay T, Muthamilarasan M, Prasad M (2017) Millets for next generation climate-smart agriculture. Front Plant Sci 8:1266
Barwant MM (2021) Plants as food source; future prospective. In: Local food and community empowerment through tourism, 2. Eureka Publications
Bocquet-Appel JP (2009) The demographic impact of the agricultural system in human history. Curr Anthropol 50(5):657–660
Buckler ES, Stevens NM (2006) Maize origins, domestication, and selection. In: Darwin's harvest: new approaches to the origins, evolution, and conservation of crops. Columbia University Press, pp 67–90
Casas A, Otero-Arnaiz A, Pérez-Negrón E, Valiente-Banuet A (2007) In situ management and domestication of plants in Mesoamerica. Ann Bot 100(5):1101–1115
Castillo C (2011) Rice in Thailand: the archaeobotanical contribution. Rice 4(3):114–120
Chaudhary J, Shelar R, Thakur K, Singh R (2023) Millets in India: production, consumption and impact on food security. Asian J Agric Extens Econom Sociol 41(8):151–162
Crymes AR (2015) The international footprint of teff: resurgence of an ancient Ethiopian grain, pp 1–10
Dijkstra A, Polman J, van Wulfften-Palthe A, Gamboa PA, van Ekris L (2008) Survey on the nutritional and health aspects of teff (Eragrostis Tef). Memorias, red-alfa lagrotech, comunidad europea, cartagena, pp 321
Dwivedi N, Rathore V, Sharma K (2023) A review of millet crops for agricultural sustainability in India. Asian J Agric Extens Econom Sociol 41(10):216–224
Ebba T (1975) T'ef (Eragrostis Tef) cultivars: morphology and classification part 2
Elliott A (2019) A mind spread out on the ground. Malahat Rev 4:7
Espitia-Rangel E, Mapes-Sánchez EC, Núñez-Colín CA, Escobedo-López D (2010) Geographical distribution of cultivated species of Amaranthus and their wild relatives in Mexico. Revista mexicana de ciencias agrícolas 1(3):427–437
Firew GA (2010) Cultivation and consumption of teff in Gojjam highlands: implication for understanding the beginning of food production in Ethiopia. Nyame Akuma 73:77–87
Fomsgaard IS, Añon MC, de la Rosa APB, Christophersen C, Dusek K, Délano-Frier J, Taberner A (2010) Adding value to holy grain: providing the key tools for the exploitation of Amaranth-the protein-rich grain of the Aztecs. Results from a Joint European-Latin American Research Project. Aarhus Universitet, Det Jordbrugsvidenskabelige Fakultet
Fornasiero A, Wing RA, Ronald P (2022) Rice domestication. Curr Biol 32(1):20–24
Geniusz MS (2015) Plants have so much to give us, all we have to do is ask: Anishinaabe botanical teachings. U of Minnesota Press
González JA, Sarmiento FO (2020) Andean indigenous foodscapes: food security and food sovereignty in mountains' sustainability scenarios. In: The Elgar companion to geography, transdisciplinarity and sustainability. Edward Elgar Publishing, pp 378–389
Grun P (1990) The evolution of cultivated potatoes. Econ Bot 44(Suppl 3):39–55
Jacobsen SE, Mujica A, Ortiz R (2003) The global potential for quinoa and other Andean crops. Food Rev Intl 19(1–2):139–148
Jancurová M, Minarovicová L, Dandar A (2009) Quinoa–a review. Czech J Food Sci 27(2):71–79
Joseph L, Turner NJ (2020) "The old foods are the new foods!": erosion and revitalization of indigenous food systems in Northwestern North America. Front Sustain Food Syst 4:596237
Joseph L (2021) Walking on our lands again: turning to culturally important plants and indigenous conceptualizations of health in a time of cultural and political resurgence. Int J Indigenous Health 16(1)
Ketema S (1997) Tef. Eragrostis tef (Zucc.) Trotter. Promoting the conservation and use of underutilized and neglected crops. 12. Institute of plant genetics and crop plant research, Gatersleben/International Plant Genetic Resources Institute, Rome
Kimmerer RW (2013) Braiding sweetgrass: indigenous wisdom, scientific knowledge and the teachings of plants. Milkweed editions

Kovach MJ, Sweeney MT, McCouch SR (2007) New insights into the history of rice domestication. Trends Genet 23(11):578–587
Lenka B, Kulkarni GU, Moharana A, Singh AP, Pradhan GS, Muduli L (2020) Millets: promising crops for climate-smart agriculture. Int J Curr Microbiol App Sci 9(11):656–668
Leonti M, Nebel S, Rivera D, Heinrich M (2006) Wild gathered food plants in the European Mediterranean: a comparative analysis. Econ Bot 60(2):130–142
Lokesh K, Dudhagara CR, Mahera AB, Kumar S, Patel HD (2022) Millets: the future smart food. Pharma Innov J 11(4):75–84
Martin C, Li J (2017) Medicine is not health care, food is health care: plant metabolic engineering, diet and human health. New Phytol 216(3):699–719
Marx JL (1977) Amaranth: a comeback for the food of the Aztecs? Science 198(4312):40
McDonell E (2020) Commercializing the "Lost Crop of the Inca": quinoa and the politics of agrobiodiversity in "traditional" crop commercialization. In: Andean foodways: pre-Columbian, colonial, and contemporary food and culture. Springer International Publishing, Cham, pp 383–406
McGrew WC (1983) Animal foods in the diets of wild chimpanzees (Pan troglodytes): why cross-cultural variation? J Ethol 1(1):46–61
Michaelraj PSJ, Shanmugam A (2013) A study on millets based cultivation and consumption in India. Int J Market Financial Services Manage Res 2(4):49–58
Mishra A (2023) Millets for sustainable agriculture. Farm Chronicle–An Agriculture Newsletter 2:10–15
Pickersgill B (2007) Domestication of plants in the Americas: insights from mendelian and molecular genetics. Ann Bot 100(5):925–940
Rashid A, Anand VK, Serwar J (2008) Less known wild edible plants used by the Gujjar tribe of district Rajouri, Jammu and Kashmir State-India. Int J Bpt 4(2):219–224
Rasingam L (2012) Ethnobotanical studies on the wild edible plants of Irula tribes of Pillur Valley, Coimbatore district, Tamil Nadu, India. Asian Pac J Trop Biomed 2(3):S1493–S1497
Raven PH (2019) Saving plants, saving ourselves. Plants, People, Planet 1(1):8–13
Rivera D, Obon C, Heinrich M, Inocencio C, Verdea A, Fajardoa J (2006) Plants–ethnobotanical investigations and historical development. Local Mediterranean Food Plants Nutraceuticals 59:18
Sangeetha J, Thangadurai D, Fayeun LS, Akinwale JA, Habeeb J, Maxim SS, Islam S (2020) Origin and evolution of rice as domesticated food crop. In: Rice research for quality improvement: genomics and genetic engineering: volume 1: breeding techniques and abiotic stress tolerance. Springer Singapore, Singapore, pp 1–14
Sani HA, Rahmat A, Ismail M, Rosli R, Endrini S (2004) Potential anticancer effect of red spinach (Amaranthus gangeticus) extract. Asia Pac J Clin Nutr 13(4):1–12
Satterfield T, Robertson LA, Turner N, Pitts A (2012) Being Gitka' a' ata: a baseline report on Gitka' a' ata way of life, a statement of cultural impacts posed by the northern gateway pipeline, and a critique of the ENGP assessment regarding cultural impacts
Schnetzler KA (2018) Food uses and amaranth product research: a comprehensive review. Amaranth Biol Chem Technol:155–184
Sedivy EJ, Wu F, Hanzawa Y (2017) Soybean domestication: the origin, genetic architecture and molecular bases. New Phytol 214(2):539–553
Smirnoff N (2011) Vitamin C: the metabolism and functions of ascorbic acid in plants. In: Advances in botanical research, vol 59. Academic Press, pp 107–177
Stallknecht GF, Schulz-Schaeffer JR (1993) Amaranth rediscovered. New crops. Wiley, New York, pp 211–218
Streetman LJ (1963) Reproduction of the lovegrasses, the genus Eragrostis. Wrightia 3(1):41–60
Sweeney M, McCouch S (2007) The complex history of the domestication of rice. Ann Bot 100(5):951–957
Talaga T (2018) All our relations: finding the path forward. House of Anansi
Tapia ME (2015) The long journey of quinoa: who wrote its history. State ART Rep Quinoa 3:1–605

Torija-Isasa ME, Matallana-González MC (2016) A historical perspective of wild plant foods in the Mediterranean area. In: Mediterranean wild edible plants: ethnobotany and food composition tables. Springer, New York, pp 3–13

Tuck E, Yang KW (2012) Decolonization is not a metaphor. Decolonization: Indigeneity Educ Soc 1(1):1–40

Turner KL, Berkes F, Turner NJ (2012) Indigenous perspectives on ecotourism development: a British Columbia case study. J Enterpris Commun: People Places Glob Econ 6(3):213–229

Turner MI, Adams KR, Berkebile JN, Dockter AR (2021) Ancient grains: new evidence for ancestral Puebloan use of domesticated Amaranth. Am Antiq 86(4):815–832

Vishwakarma KL, Dubey V (2011) Nutritional analysis of indigenous wild edible herbs used in eastern Chhattisgarh, India. Emir J Food Agric 23(6):554–560

Willett W, Rockström J, Loken B, Springmann M, Lang T, Vermeulen S, Garnett T, Tilman D, DeClerck F, Wood A, Jonell M (2019) Food in the Anthropocene: the EAT–lancet commission on healthy diets from sustainable food systems. Lancet 393(10170):447–492

Winkel T, Aguirre MG, Arizio CM, Aschero CA, Babot MDP, Benoit L, Joffre R (2018) Discontinuities in quinoa biodiversity in the dry Andes: an 18-century perspective based on allelic genotyping. PLoS One 13(12):e0207519

Xue S, Bradbury PJ, Casstevens T, Holland JB (2016) Genetic architecture of domestication-related traits in maize. Genetics 204(1):99–113

Zohary D (2017) The progenitors of wheat and barley in relation to domestication and agricultural dispersal in the Old World. In: The domestication and exploitation of plants and animals. Routledge, pp 47–66

# Chapter 2
# Uncommon Edibles: Rediscovering Forgotten Plants

**Abstract** Hunger, malnutrition, poor health, and even starvation continue to be the biggest problems facing the world. Malnutrition is characterized as a lack of nutrients brought on by either eating insufficient amounts of food or consuming food that is low in nutrients compared to the daily requirements. The low-income families frequently cannot afford animal-based diets, a variety of uncommon edibles or neglected crops can provide an alternate supply of vitamins, minerals, and secondary plant compounds that are beneficial to health. Therefore, studies in horticulture and agriculture ought to develop plans to increase access to nutrient-dense food as well as to generate more food. Although many different plant species are designated for consumption as fruits, vegetables, and grains worldwide, only a small percentage of these species are utilized as commercial crops. As a result, many uncommon, underutilized and neglected species present opportunities to enhance food production and diversify the human diet, hence facilitating more resilient and sustainable horticultural and agro-food systems. Therefore, the current book chapter highlights the significance of uncommon edible species as Future Food for enhancing human nutritional status and boosting the resilience of horticultural and agro-food systems.

**Keywords** Andean Roots and Tubers · Hunger · Uncommon Edibles · Malnutrition · Neglected Plants

## 2.1 Introduction

Uncommon edibles are fruits, vegetables, grains, and other "ingredients" that are no longer in trend and don't make up significant amount of the cuisines of today. The majority of participants in the global food systems frequently ignore the function of uncommon edibles also known as forgotten crops, although certain food and nutrition security companies, like the Food and Agriculture Organization (FAO), have acknowledged their potential. Adding "uncommon edibles" to the global food system could increase the risks of food insecurity and create more robust and healthful



N. Patel et al., *The Future of Food: Exploring Edible Plants for a Sustainable Diet*,
https://doi.org/10.1007/978-3-032-15027-1_2

food systems for coming generations, but there are still numerous obstacles in the way (Ingram and Zurek 2019; Ulian et al. 2019).

It can be challenging to define uncommon edibles, but the most important thing to keep in mind is their small contribution to larger food systems. Crop species that are no longer used commercially as well as cultivars that have been superseded by improved cultivars, as well as their wild relatives and landraces, can all be considered forgotten crops/Uncommon Edibles (Hammer and Teklu 2008). Crop landraces are distinct because they are dynamic populations of cultivated plants with a particular identity, historical origin and lack formal crop improvement. They are also sometimes genetically diverse, naturally adapted and linked to traditional farming systems, which makes their use promising (Lim 2012).

The majority of uncommon edibles in the UK are officially preserved by a number of seedbanks, gene banks, and research organizations like the Millennium Seed Bank, John Innes Centre and Science and Advice for Scottish Agriculture. These accessions frequently span in the thousands, as is the case with the Germplasm Research Unit-maintained undervalued wheat cultivars (*Triticum aestivum* L.) or pea cultivars (*Pisum sativum* L.) (Villa et al. 2005). Although some of these underutilized cultivars are actively employed on a small scale, their influence is restricted to supporting the local food chain and playing ancillary roles (Hammer and Teklu 2008).

A robust, resilient, and long-lasting food and nutrition security system might be established with the help of uncommon edibles and forgotten landraces. A partial shift to less well-known crops, like some of the fruits and vegetables, might boost domestic production, support small and medium-sized farms, provide employment and bring back the lost cultural and gastronomic traditions even in the United Kingdom (Padulosi et al. 2002). Furthermore, these crops have the potential to boost agrobiodiversity because they are less genetically developed, which makes them more adaptable to intercropping and propagation in less intensive, more conventional agricultural systems that require fewer chemical treatments. However, many traits of the majority of the overlooked crops' are still not fully understood.

## 2.2 Why There Is a Need to Discover Uncommon Edibles?

Beyond nutrition, uncommon edibles excel in sustainability due to their higher nutritional diversity and lower glycemic index. They can be cultivated in urban or marginal regions with less water, fertilizer, and/or pesticide, while enhancing soil fertility. Seed banks, community gardens, and internet resources are increasingly being used to address some of the issues of limited availability or even how to know what one should look for. Through agricultural diversification, they lower the risks of crop failure and market instability. By generating value-added products, the production of these crops in rural areas boosts local economies and reduces dependency on imports.

## 2.3 Uncommon Wild Greens and Leafy Vegetables

### *2.3.1* Urtica dioica *L.*

Many scientific studies have focused on identifying the components of inexpensive herbs that may be useful in enhancing human health in the past few decades. The fact that almost no uncommon, exotic or beautiful plant has proven more fascinating has been a startling aspect of herbal study. Generally, it has been found that certain common, unfamiliar and despised weeds contain unimaginable benefits (Thornhill and Kelly 2000). *Urtica dioica* is a weed that has medicinal uses for its leaves, seeds and even roots. It is a reasonable storehouse of dietary components and micronutrients. Despite being healthy, leaves are abundant in pro-vitamin A, vitamin C, and minerals, including iron. They are also readily digested. Due to its vast distribution and remarkable biological properties, *U. dioica* is utilized as food and medicine in many nations, especially in the Mediterranean region (Hartl and Vogl 2002).

*Urtica dioica* also known as stinging nettle is indigenous to Europe, Asia, America and Northern Africa. It belongs to Urticaceae family. This perennial flowering herbaceous plant has a diverse record of use as a herbal medicine and dietary supplement. *U. dioica* has been used for decades to treat multiple conditions, from hypertension to allergic rhinitis (Jan et al. 2017). Apart from being a blood-nourishing tonic, it can be used as a tea, vegetable, and juice, mostly for rhinitis. One important feature of *U. diocia* is that the entire plant may be employed for a number of purposes, such as food, medicinal, fodder, textile processing, biodynamic agriculture and cosmetics (Rutto et al. 2013). The tender leaves can be dried for use in the winter and added to soups. Its use as an edible herb/green leafy vegetable is not traditional, rather, it was taught by European traders and settlers, or perhaps by Chinese immigration in some areas. However, it quickly gained popularity as a cooked green and was utilized by the Iroquois as well as a variety of indigenous peoples in the west, such as the Haida, Okanagan Colville, Nlaka'pamux, Lillooet, Halkomelem, Sechelt, Straits Salish, and Tanaina of Alaska (Taylor 2009; Halder and Sharma 2017).

### *2.3.2* Aegopodium podagraria *L.*

*Aegopodium podagraria* L. also referred as goutweed is regarded as an uncommon leafy vegetable used in European cuisine and for foraging. In Poland, this is the only species of *Aegopodium genus*. Originating from the Greek words aigos, aix and podion, the Latin name of this species describes the arrangement and form of the foliage segments, which mimic the form of a goat's hoof (Kyrbassova et al. 2018). Both goutweed and podagraria refers to the function of plant, as it was employed in arthritis treatment during the Middle Ages. Invasive goutweed is primarily found in the moderate zone of northern hemisphere. Although it is indigenous to Europe,

Siberia, the Caucasus, Kazakhstan, and Central Asia, it is currently spreading to North America, Australia and Japan. Its main habitats are gardens, ditches, thickets, and shaded, damp woodlands (Valyova et al. 2016).

With a long growing season, this plant serves as ground cover that spreads widely and appears early in the spring. The site, soil fertility and irrigation all affect the height of plant (60–90 cm). Leaves, fruits, seeds, rhizomes and flowers are among the parts that are utilized while april and may are ideal for leaf harvesting (Orav et al. 2010; Sayed-Ahmad et al. 2017).

The chemical composition of this plant is extremely diverse, containing both primary and secondary metabolites that are produced for specific purposes, such as protection against harmful environmental elements (pests, UV radiation). It contains lectins, glycoprotein-binding carbohydrates and carbohydrates like fructose and glucose (Dębia et al. 2025). These compounds can have positive health benefits on the human body and are a great source of nutrients. Goutweed use can offer beneficial nutrient supplement of antioxidant components that are also present in more widely available foods like fruits and vegetables. In addition to phytocompounds, the leaf portion supply vitamin C, lutein, neoxanthin, violaxanthin, antheraxanthin, and zeaxanthin carotene (Augspole et al. 2017; Savinov et al. 2019).

### *2.3.3* Smyrnium olusatrum *L.*

*Smyrnium olusatrum* L., which is part of the Apiaceae family, is referred to as Macerone, "Corinoli, or Sedano selvatico in Italian, Alexanders" and Wild celery in English, and Maceron in French. This sturdy, glabrous, celery-scented biennial herb grows up to 150 cm in height and blooms just once before withering. Because of the distinct scent of the fruits, which exude a strong myrrh-like oil, the name stems from the Greek word rltqma, which signifies myrrh. The species' name originates from the Latin word olus atrum, which means "black herb," referring to the color of the ripe fruits, which are 7–8 mm schizocarps made up of two single-seeded mericarps apiece (Brinkkemper 2015; Messaoud et al. 2013).

This plant has been utilized as a culinary vegetable since ancient times (such as the Roman era) due to the aromatic flavor of its parts. Additional evidence suggests that the Romans introduced this plant to Britain for both culinary and therapeutic purposes. Thus, the plant was grown throughout Europe until the Middle Ages, when celery (*Apium graveolens* L.) usurped its place (Hamdi et al. 2025). The entire plant can be consumed: the green leaves are used as a vegetable or to make a white sauce; the tender stems are cooked like asparagus; the flower buds are included in salads or fritters or even steamed and eaten in place of broccoli; the fruits are used to season meat, soups, and salads and as a pepper substitute; and the raw or cooked roots are served at the table, frequently in place of parsnips (Heywood 1972). While it is widespread in the northern region, primarily in Emilia-Romagna, *S. olusatrum* is found in central and southern Italy in addition to the islands. The particular species is found throughout Europe and beyond, from nations that border the

Mediterranean to Scotland and Ireland, which are at the northernmost point of the continent and have completely naturalized it following its introduction during the Roman era (Randall 2003; Petrelli et al. 2017).

### *2.3.4* Moringa oleifera *Lam.*

A significant portion of the population is undernourished, which means they lack access to enough food. This leads to a high rate of malnutrition and other nutritional impairments. Growing natural resources that are used in some nations and underutilized in others could provide important nutrients that are accessible to a large portion of the population and help address the malnutrition issue (Patel et al. 2023). The most well-known and widely used species in the tropics and subtropics, *Moringa* is a member of the Moringaceae family and is native to the sub-Himalayan region of India, Bangladesh, Pakistan and Afghanistan. It is also native to several other regions where it is known as the miracle/drumstick tree due to the shape of its pods (Patel and Krishnamurthy 2023).

Green leafy vegetables including African spinach, amaranth, leafy eggplant, tossa, and drumstick (*M. oleifera*) leaves are grown year-round in Ghana and can be consumed on a daily basis. Among Ghana's rural districts is Ada-East district the indigenous people are primarily farmers and fishers. The majority of them grow cassava, tomatoes, pepper, *Moringa* and onions. As green leafy vegetable, *Moringa* is a great source of folic acid, vitamin C, provitamin A, carotenoids and nutritional minerals like manganese, calcium, zinc, and iron. Provitamin A carotenoids, vitamin C, folic acid, and dietary minerals like iron, calcium, zinc and manganese are all abundant in *Moringa* plants. Despite having excellent nutritional and phytomedicinal characteristics, *Moringa* remains rarely used in some parts of the world (Patel and Krishnamurthy 2021; Patel et al. 2025).

### *2.3.5* Houttuynia cordata *Thunb*

*Houttuynia cordata* Thunb., also known as Yuxingcao in Chinese, is a significant vegetable and therapeutic plant that is consumed in southern China. Originating from China, Japan, Korea, northeast India, and Southeast Asia, *H. cordata* is a perennial flowering herb that is polyploid (diploid or tetraploid) (Ma and Wu 2003). This plant has a distinct fishy scent, blooms white flowers in the early summer, and grows well in moist to soggy soil on shaded slopes, waysides, and ridges of fields between 300 and 2600 meters above sea level. Remarkably, this herb does not reproduce sexually; instead, it spreads by parthenogenesis, development and separation of underground stems (Kwon and Ha 2012). *H. cordata* is an essential herb plant resource for the food, pharmaceutical, and aromatic industries. It is a common aromatic, medicinal and vegetable plant.

Due to its high nutritional and economic importance, *H. cordata* is a vegetable that is abundant in proteins, minerals, vitamins, and amino acids. There exists ample evidence of *H. cordata's* traditional or ancestral knowledge. The scientific knowledge of *H. cordata* and other medicinal plants was unknown to the people of the past, but they did have traditional knowledge that was passed down from one generation to the next through ancestral therapists, knowledgeable elders or ordinary people in the absence of any relevant records. This species, *H. cordata*, is used by humans to satisfy a variety of economic purposes (Yu et al. 2022). Different communities and tribes in different locations have distinct names for this species and was used in a variety of ways.

Several parts of this plant are used by the Maonan and Hakka and Hokles people in Guangdong province, China, to make herbal soups, teas, baths, and decoctions. They utilize it to treat a number of illnesses, including diarrhea, dysentery, flu, cough, sore throat, and enteritis. It is used by the people of Nujiang, North-West Yunnan, China, to treat snakebites, tonsillitis, nephritis, and enteritis (Nguanchoo et al. 2014). In Thailand, this plant is referred to as "Pak Kan Thong" a popular vegetable among those living in the country's eastern and northern provinces. For decades, numerous cultures around the world have used this plant species for a variety of uses based on their needs and requirements (Pradhan et al. 2023).

### *2.3.6* Alternanthera sessilis *(L.) R. Br. ex DC*

The plant *Alternanthera sessilis* (L.) R.Br. ex DC., commonly referred to as sessile joyweed, dwarf copper leaf, or mukunuwenna in Sri Lanka, is indigenous to Brazil and can be identified in many tropical and subtropical areas across the world. Although it can grow in a range of soil types, this perennial plant thrives in damp environments and can be found in both marshes and uplands. The plant is a glabrous, succulent herb that stands upright, sometimes branching, and can reach a height of one meter (ole Alwis et al. 2006).

This plant has certain advantageous traits while being a staple food in Sri Lanka. Despite having a more herb-like appearance than popular leafy vegetables like spinach or cabbage, it is classified as a leafy vegetable. Known by many as "mukunuwenna," it is the most popular and commonly grown of the nine leafy vegetables grown and marketed in Sri Lanka (Lin et al. 1994). Once grown, it may be harvested monthly for roughly 2 years, is simple and affordable to nurture and sustains little harm during transportation. Sri Lankans eat the plant multiple times a week because they think it has high vitamin, protein, and fiber content. Apart from Sri Lanka, the leaves and stems of *A. sessilis* are consumed as a cooked vegetable in many other parts of the world. It has been used as a vegetable in tropical Africa including Guinea, Benin (in soups and sauces), Nigeria (in soup), Congo, Tanzania, and Zambia (as a relish), Madagascar, and Reunion (as a pot herb). *Alternanthera sessilis* is used as a plaster for sick or injured skin, as well as for basic stomach issues, diarrhea, and dysentery. The crushed plant is used to treat vertigo and headaches in Nigeria. Leafy twigs that have been grounded into a powder are used to treat snakebite in Senegal and India (Gunasekera 2008).

### *2.3.7* Cakile maritima *Scop*

*Cakile maritima* Scop., or European Sea rocket is annual species of the Brassicaceae family that is found throughout the coasts of all five continents in Europe. This plant can cover regions up to one meter in diameter, are heavily branched, and are no taller than 40 cm. The succulent, glabrous leaves can have pinnate lobed, oblate, or oblanceolate shapes. The flowers are very fragrant, having four white to purple petals, and can reach a diameter of up to 25 mm (Gandour et al. 2008). While self-pollination is uncommon, cross-pollination is typically mediated by several insect species.

The wild species *Cakile maritima* is extensively distributed and has an ecologically significant role in coastal areas. Individuals harvested from their natural habitats have historically been used as culinary or therapeutic plants, despite the species' lack of commercial cultivation. In the current climate change scenario, farmers, markets, and consumers are shifting back to stress-tolerant wild plants like halophytes, including *C. maritima*, or resilient traditional species and cultivars as a base for recovered or new cash crops that are better suited to the difficult climatic conditions (Daniela et al. 2010; Conversa et al. 2024).

The recipes prepared by using roots, young leaves, stems, flowers and fruits are available, and almost all of the plant parts are edible (Ksouri et al. 2007). Sea rocket stems and young leaves make excellent additions to fresh salads, but older leaves are suggested for use in soups, as a potherb, or to extract flavourings for salsas or other dishes because of their harsh taste. Similarly, fresh flowers can enhance soups with a hint of bitterness and are a gourmet addition to fresh salads (Meot-Duros et al. 2008).

### *2.3.8* Adansonia digitata *L*

*Adansonia digitata* L is one of the most well-known members of the small pantropical tree family Malvaceae. It is commonly known as African Baobab. This tree is more commonly recognized as a fruit, fiber and a vegetable tree. According to recent studies, baobab leaves are consumed all over its range, but particularly in West Africa (Gebauer et al. 2002). This species has a wide range of distribution, occurring in western Madagascar and most of semi-arid and sub-humid parts of Sub-Saharan Africa. It has been successfully introduced and grown outside of Africa (Iyaka et al. 2014).

Many people in Africa, particularly those in the central region, rely on the leaves of the baobab tree as a primary food. Fresh leaves are harvested during the monsoon season when they are at their tender stage dried for use at home and for sale during the dry season. The leaves are used as an immunological stimulant and as a febrifuge. They include vitamins, minerals, and bioactive compounds. Usually, the leaves are pounded and sieved into a fine powder or sun-dried and preserved whole. Young leaves are often utilized, cooked like spinach, dried, and powdered and incorporated into porridge sauces, thick grain gruels, or boiled rice (Catarino et al. 2019).

## 2.4 Grains, Seeds and Tubers as Uncommon Edibles

### *2.4.1* Fagopyrum esculentum *Moench*

In the majority of places, *Fagopyrum esculentum* Moench also referred as buckwheat is a fairly new crop. Buckwheat was first grown in China slightly earlier than the second century B.C., and it wasn't until the late Middle Ages that it made an appearance across India, Russia, and Europe, despite perhaps being grown in Japan as early as 3000 years ago (Zhou et al. 2012). Although the crop is not true cereal, the seeds (achenes) are typically categorized as such due to their similar applications. The grain is typically fed to animals and consumed by humans. The flour can be used to make pancakes, biscuits, noodles, cereals, and more. The dehulled groats are suitable for use as porridge (Ahmed et al. 2014).

Buckwheat has a substantial amount of fiber, zinc, copper, magnesium, and selenium, as well as proteins with high biological value and balanced amino acids. The China, Russian Federation, Kazakhstan and Ukraine are currently the top producers, with production occurring in other nations on different continents. Europe and Asia are the two continents with the most production (Guo et al. 2010). Buckwheat comes in two primary varieties: Tartary buckwheat, which is bitter, and ordinary/common buckwheat, which is sweet. Due to its superior palatability, consumers typically favour common buckwheat over tartary buckwheat. A pseudocereal, common buckwheat is used in a manner analogous to that of wheat. Because of the triangular shape of its seeds, resembling seeds of the beech tree, the term "buckwheat" stems from the words buck and wheat, suggesting its similar use as wheat (Luitel et al. 2021). Carbohydrates, which are biochemically comparable to cereal starch, make up 60–70% of buckwheat grains. In addition, the starch granules may include bioactive substances including trypsin inhibitors, anticancer proteins and hypotensive peptides, as well as nutrients like lipids, and carbs (Wang et al. 2007; Shaheen et al. 2025). There are several health and nutraceutical advantages of buckwheat. Because of its inability to adapt as per farmers' point of view, it remained an overlooked crop, but growing public knowledge continues to make it becoming increasingly common.

### *2.4.2* **Triticum turgidum** ***ssp. durum***

Some *Triticum* species were domesticated from wild predecessors that grew in the Fertile Crescent at the onset of agriculture, in the late Stone Age. Among the first wheat species to be cultivated were emmer (*T. turgidum* L. subsp. *dicoccum* (Schrank) Thell) and einkorn (*Triticum monococcum* (L.), which were staple crops for several millennia. Common (*T. aestivum* subsp. *aestivum* (L.) and durum (*T. turgidum* L. subsp. durum (Desf.) Husnot) wheat, though spelt (*T. aestivum* subsp. spelta (L.) Thell), another hulled wheat species, have gradually supplanted these hulled wheat species with more productive and free-threshing species (Bordoni et al. 2017).

For over 8000 years, wheat has been the fundamental staple diet of the main civilizations in West Asia, North Africa and Europe. It was the earliest crops to be domesticated. Wheat is currently the most important staple food for humans and is cultivated on more acreage than any other commercial crop (Molberg et al. 2005).

The term ancient wheat refers to wheat that was utilized by prehistoric cultures. Einkorn, Emmer, Khorasan, and Spelt are typically regarded as ancient wheat. Heritage wheat is another word for wheat that is widely cultivated during the transition from ancient to modern wheat. This wheat is made up of types that were chosen from either wild or ancient wheat. Land races, which were composed of numerous closely related strains, constitute the majority of ancient and heritage wheat (Wiwart et al. 2013). Most of these ancient grains, which have not undergone significant genetic advancements, have been reintroduced in the past 10 years, and interest in alternative cereals has grown as people become more conscious of what foods are deemed natural and healthful. This interest is also linked to some of them that are said to be more tolerable by people who have modern wheat allergies or intolerances (Piergiovanni et al. 2009). When assessing the potential of an ancient crop species like Oriental wheat, it is important to keep in mind that the ever-tightening quality standards of contemporary food industry should not be exaggerated. The ability of these species to replicate traits found in contemporary wheat cultivars is less important from an economic standpoint than their potential for application development of innovative wheat-based foods.

### *2.4.3* **Digitaria exilis Stapf** ***and*** **Digitaria iburua Stapf**

*Digitaria exilis* Stapf and *Digitaria iburua* Stapf, also referred to as fonio, acha rice, hungry rice, or hungry millet are members of the Poaceae family's Genus Digitaria, which is subfamily Panicoideae. Living in middle Niger, the Mande language group is the source of the colloquial term "fonio" or "fonyo." (Adoukonou-Sagbadja et al. 2006). About 230 different kinds of fonio, an ancient African crop, are grown on several hundred thousand hectares of cropland throughout the tropics and subtropics. West Africans have been growing it over the arid savannas for more than thousands of years (Nyam et al. 2017).

Fonio offers rural residents an inexpensive supply of nutritious food and high-quality, nutrient-dense animal feed. Fonio is consumed two to three times a day in some parts of West Africa, where it is preferred above other grains. Despite being served with affection to dignitaries during special events, it is regarded as being extremely healthy and serves a significant part in household food security (Animasaun et al. 2018). Additionally, fonio chaff is frequently used as cattle feed, and the grain is used to feed chickens. Fonio is traditionally advised as a diet for diabetics and nursing mothers. In addition to being eaten as food, the straw is a valuable component of local roofs and mattresses, and the grain is utilized in Dominican Republic religious ceremonies (Ballogou et al. 2013).

Methionine and cystine, two essential amino acids that are typically lacking in major grains, constitute 7.3% of the fonio millet protein. Additionally, fonio millets have more amino acids than whole-egg protein (12%) for leucine and (18%) for methionine, with the exception of lysine, which is lower at 46%. Because fonio has a low natural sugar content, it is safe for human consumption (Bassey et al. 2023). Some phenolic chemicals and their derivatives, including gallic acid, ferulic acid, caffeic acid, sinapic acid, and vanillic acid, are present in the grains. Even though the crop is important to West African traditional agriculture, especially in Nigeria, there is currently a lack of study on how to improve it. Consequently, the crop continues to be archaic due to a number of agronomic problems, including low yield, small grain size, seed cracking, plant lodging, fugacious and shattering seeds among others, which reduces the yield of grains (Koreissi-Dembélé et al. 2013; Animasaun et al. 2023).

### 2.4.4 Tylosema esculentum *(Burch.) A. Schreib*

The *Tylosema esculentum* (Burch.) A. Schreib also known as Marama bean is sometimes referred to as the "green gold," has been claimed to be indigenous to the South African Kalahari Desert and is also grown in Namibia and Botswana. Australia, Israel, Kenya, and the United States of America are currently performing experimental cultivation of the marama bean. Zambia and Mozambique additionally cultivate this plant in small quantities. The South African provinces of the Northern Cape, Limpopo, North West and Gauteng are native to marama beans. Legumes are outstanding resources for ensuring food and nutrition security worldwide since they are vital source of protein, fiber, vitamins and minerals (Omotayo and Aremu 2021). It is crucial to consume beans as a less costly option in developing nations where protein-energy deficiency is frequently a serious nutritional issue. Native legumes have the capacity to boost diet protein content in underdeveloped nations. However, the potential nutritional and economic benefit of many uncommon indigenous legumes is limited since they are not domesticated. The current reliance on a few number of legumes that may bring agronomic, nutritional, economic, and ecological risks can be lessened by investigating the domestication and cultivation of overlooked uncommon plants, such as marama beans. Effective domestication, cultivation, preservation, and conservation of native plants are crucial for the exploration of their nutritional and commercial potential (Holse et al. 2010).

Marama is a substantial source of a variety of nutritional and phytochemical substances. Recent publications in the literature have detailed the chemical nature of marama beans and compared them to other legumes high in protein. Its phytochemical and nutritional content has a wide range of applications in the pharmaceutical, nutraceutical, and functional food industries. Compared to common staple tubers like potatoes, yams, and sugar beets, the underground portion of marama yields a larger and more nutrient-dense tuber. The tuber can be eaten raw, cooked, or roasted, but it must be collected when the plant is still young, about 1 or 2 years old, when it is still succulent and nutrient-dense enough to eat (Adeboye and Emmambux 2021).

Although the tuber turns fibrous and bitter, it may be a valuable source of water for farmers, hunters, and animals in an emergency. When fully grown, this tuber can weigh 12 kg and have an average water content of 90% (Jackson et al. 2010). In addition, the tubers contain an enormous amount of carbohydrates. The texture of young marama tubers are frequently compared to that of artichokes, and they have a pleasant, succulent, and sweet taste. When dried and edible, marama tubers have a crimson color. Furthermore, the mature and immature stems can be used to make soup and are edible. Additionally, it is thought to be exceptionally high-quality livestock feed (Hamunyela et al. 2020).

### *2.4.5* Plukenetia volubilis *L.*

The Amazon basin is home to the underused uncommon edible oilseed crop *Plukenetia volubilis* (Euphorbiaceae Family), which has been employed by people since the Incan era. Of the 45–50% lipid content of the bold seeds, about 35–50% is α-linolenic acid (and about 33–41.% is linoleic acid, the two important fatty acids that humans need. Additionally, the seeds have antioxidant qualities and contain 22–30% protein. It has received more attention lately and is being grown more widely because of its superior nutritive profile and favorable agronomic qualities (Loaiza 2013). In light of today's global issues, a reform of our food systems is crucial for ensuring food security, reducing the effects of climate change and malnutrition.

*Plukenetia volubilis* has been connected to mankind from pre-Hispanic times and has long been consumed in Latin America. Incan burial sites along the Peruvian coast have yielded artifacts that shows the fruits and vines of this plant, suggesting that it may have been grown by the Incans between 3000 and 5000 years ago. *P. volubilis* and other large-seeded species in the genus are most frequently referred to by their Quechuan colloquial name, "sacha inchi." But depending on the dialect, Sacha inchic or Sacha inchik are also used, and they have the same meaning. Inchi indicates groundnut or peanut, while sacha can mean mountain, false, or resembling (Rodzi and Lee 2022).

*Sacha inchi* is an underappreciated/uncommon edible native plant that is used in traditional medicine. It is also showing enormous promise for facilitating the development of innovative pharmaceutical and nutraceutical products with added value (Kodahl and Sørensen 2021) Laboratory, preclinical, and clinical studies have provided strong evidence for the plant's use in macro/micronutrients, α-linolenic acid, phytochemical extraction, organoleptic enhancement, neuroprotection, dermatology, antidyslipidaemic, antioxidant, and anti-inflammatory activities. The vast concentration range continues to be a barrier for both medication and nutraceutical evaluation, even with the encouraging outcomes of the advanced studies (Wang et al. 2018; Kodahl 2020).

### 2.4.6 Andean Roots and Tubers

In Andean upland populations, tubers and roots are a staple diet that are prized for their high carbohydrate content, of which starch accounts for the majority. Consuming them is also recommended for a number of health advantages, such as reducing pain, wound healing, diabetic control, digestive system enhancement, and anti-inflammatory properties (Flores et al. 2003). Numerous studies on Andean tubers, including *Smallanthus sonchifolius, Arracacia xanthorrhiza, Ullucus tuberosus, Pachyrhizus ahipa, Tropaeolum tuberosum*, and *Oxalis tuberosa*, have identified many compounds that have pharmacological effects on people and boost immunity (Leidi et al. 2018).

*Arracacia xanthorrhiza* also known as arracacha is a tuberous root that has long been used to make bread, cakes, and beverages as well as to make soups, stews, and purees. It is widely used for feeding infants, the elderly, and those recovering from ailment in a number of South American nations, including Bolivia, Brazil, Colombia, Peru, Ecuador, and Venezuela (Choquechambi et al. 2019). It contains vitamins A, B, and C, but its high starch content renders it primarily an energy source. Its high Ca, P, and Fe content has also been reported. It is commonly believed that arracacha provides therapeutic benefits for a number of applications, including aiding with placental removal after delivery (Lim 2014).

*Pachyrhizus ahipa* also known by the name ahipa also produces tuberous roots, and it is primarily eaten fresh as a fruit but can also be used as a gluten-free flour substitute. Compared to arracacha, these tubers have a higher crude protein content. It has a crispy texture yet may be eaten fresh, peeled like a banana. Its consumption is based on the root's pleasant flavor and the general perception of its therapeutic benefits (excellent for the lungs, cures throat infections) (Padulosi et al. 2014).

The flavorful and fleshy roots of *Smallanthus sonchifolius*, commonly referred to as yacon, possess a high concentration of fructooligosaccharides, fructose, and water. Yacon is farmed mostly for the consumption of villagers in the Andes, from North Argentina to Ecuador. Although it has historically been used as a fruit, it is now used to make ice cream, cake, yoghurt, and syrups. In Japan and New Zealand, where it is grown professionally for food or feed, it is gaining popularity due to its bioactive ingredients. Although the raw root is frequently eaten as fruit, its traditional use suggests it for diabetics, those with kidney and digestive issues, and even provides some skin-rejuvenating properties (Espín et al. 2004).

The root crop *Tropaeolum tuberosum*, also referred to as mashua, may be a source of bioactive substances with advantageous properties. It is grown in Bolivia and Colombia. In addition to its high sugar and carbohydrate content, the tuberous root contains various phytocompounds (glucosinolates, antioxidants) which provide considerable medicinal potential and are being studied scientifically. Chroniclers from the fifteenth century claim that the Incas gave their soldiers mashua to help them forget about their women while they were engaged in combat (Aruquipa et al. 2016).

*Ullucus tuberosus* also referred as Ulluku, is a Quechua word, which is the source of the name Ulluco. There are two subspecies of Ulluco: tuberosus, consisting of all cultivated ullucos, and aborigineus, encompassing all wild forms. Tuberosus may have descended from the later species. The plant favors organic matter-rich soils. When ulluco is manufactured in large quantities, it is dehydrated to yield a product that retains for years. Ulluco is typically more expensive than potatoes in Andean rural and urban markets, and it seems to have enough variety to be grown in places other than the Andes, such the world's temperate zones (Busch and Savage 2000). On the other hand, *Oxalis tuberosa* also known as Ok'a, occa, and uqa are the origin of the name oca. The upper portion of this perennial plant oca senesces, but the tubers remain to spur new growth the next season. Oca is farmed on about 21,000 to 22,000 hectares of land in Peru in sandy soils, producing up to 25–40 tonnes per hectare. Oca is cultivated in Mexico and New Zealand in addition to the Andes. These tubers and roots hold a great deal of prospective to develop into novel and significant nutraceutical sources that could enhance human health (Zhu and Cui 2020). Additionally, they may be useful to other sectors. For instance, rotenone, a potent insecticide in organic farming, exists in ahipa seeds; chemical compounds with nematicide, insecticide, fungicide, and bactericide qualities have been identified in mashua and yacon. Additionally, the culinary and cosmetic industries require the strong and stable color pigments that exist in oca, ulluco, and mashua tubers (Roca et al. 2007).

### *2.4.7* Pachyrhizus erosus *L. Urban*

*Pachyrhizus erosus* L. Urban also known as Mexican yam bean or Jicama is one underappreciated/uncommon edible crop that merits consideration. It is a leguminous root that is high in nutrients, grows well in a range of temperatures, and has promise for application in industry and food. It is still a crop that is mostly grown in conventional producing regions, despite its advantages (Chung et al. 2020). On a worldwide level, its limited market recognition has prevented it from realizing its full potential. However, if given the opportunity, this robust, economically viable crop might play a significant role in increasing food security and diversifying diets.

This neglected leguminous root crop, which is native to Mexico and Central America's semiarid tropics, is a member of the Fabaceae (Papilionaceae) subfamily. Warm, humid, tropical settings are ideal for the plant's robust climbing or trailing growth. In Central America, China, India, Bangladesh, Southeast Asia, Indonesia, the Caribbean, French Guiana, Brazil, and Central and West Africa, it is grown mostly for its crunchy and luscious storage roots (Lukitaningsih 2014).

Rotenone, a naturally occurring plant-derived pesticide that is frequently employed in organic farming and pest control, is abundant in Mexican yam bean seeds. Rotenone is a potential biopesticide substitute for synthetic chemicals since it, together with other rotenoids and isoflavonoids, has strong insecticidal and acaricidal activity against thrips, aphids, and white larvae. Compounds from Mexican yam beans have intriguing potential for a variety of industrial uses due to their pharmacological, cosmetic, and biopesticidal properties, making them a useful underutilized crop for sustainable development (Ravi et al. 2025).

### *2.4.8* Phaseolus acutifolius

*Phaseolus acutifolius* also referred to as Tepary beans are indigenous to Mexico, Central America, and the southwestern United States. They thrive in the hot, arid climate of the western portion of this region. Wild teparies were collected for human food throughout this region, however this custom declined in the early 1900s (Schinkel and Gepts 1988). Cultivated teparies in the Desert region of Sonoran were traditionally irrigated with seasonal flood waters and planted with pinto beans, sorghum, and corn. As a significant subsistence crop throughout its natural habitat, cultivated tepary holds cultural significance for tribes like the Tohono O'odham, whose folklore describes white tepary beans strewn across the darkness of sky (Mhlaba et al. 2018).

Tepary bean production peaked at the start of the twentieth century, but war, shifting consumer preferences, and the advent of mechanical irrigation caused their popularity to rapidly decline. Tepary beans have a firm, sweet flavor. They can be incorporated into soups, chilis, and other dishes, or they can be prepared as pot beans (Porch et al. 2017). Tepary beans were deemed acceptable or even preferable to cowpea and fava beans used in local recipes by a tasting panel assessing their appropriateness for various Middle Eastern or African cuisines (Rao et al. 2013).

Since teparies are more resilient to abiotic stresses than most other pulses, this uncommon edible bean is receiving much-needed attention. Teparies perform better than other ordinary beans under extreme circumstances, according to yield assessments conducted under a variety of growth environments. Numerous characteristics, including as canopy biomass, pod partitioning, cellular processes, and more, are linked to this resistance to heat and drought. Teparies are viewed by several researchers as a possible genetic source of drought and heat resistance to enhance the commercially significant common bean (Moghaddam et al. 2021).

## 2.5 Uncommon Edible Fruits

### *2.5.1* Plinia cauliflora *(Mart.) Kausel*

In advanced and developing nations, there has been a growing interest in edible tropical fruits because of their possible health advantages. In Argentina and Paraguay, *P. cauliflora* also referred to as Jabuticaba is common in open woodlands and alluvial plains. The northeast, central-west, southeast, and south regions of Brazil are native to it. Although the species is widespread in the Caatinga, Cerrado, and Pantanal biomes, it is most frequently found in the Atlantic rainforest. It produces sweet, delicious fruits with a hint of acidity (Crozier et al. 2009). They can be used to make juice, jams, liquors, and wines, or they can be eaten raw.

Its sun-dried rind has long been used to treat respiratory issues like hemotypsis, asthma and chronic tonsil inflammation, as well as diarrhea, with an astringent decoction. Numerous conventional nutrients including vitamins, minerals, carbohydrates, salts and amino acids, are prevalent in fresh edible fruits of this plant (Silva et al. 2008). While some of these nutrients are found in little amounts, they can have a big impact on human health. Significant amounts of specific amino acids, such as lysine and tryptophan, are also present in jaboticaba. This fruit contains trace amounts of vitamins B1 and B2. Ascorbic acid and vital water-soluble antioxidant ingredient (Wu et al. 2013; Junior et al. 2019).

### *2.5.2* Asminia triloba *(L.) Dunal*

In 1541, a Portuguese official who traveled with Spaniard Hernando de Soto in the Southeast United States wrote the first documented account of *Asminia triloba* also known as pawpaw. He observed that throughout the Mississippi Valley, Native Americans grew and consumed pawpaws. *Asimina* fossils date back to the Eocene, which occurred roughly 56 million years ago, and the Miocene, which occurred roughly 5.3 million years ago, when the first fossils that resembled *A. triloba* appeared (Hormaza 2014). The only plant in the Asimina genus that yields fruits that are particularly valuable as food is the pawpaw. It is actually a giant fruit that is native to North America and may be eaten. In hardwood woods with damp but well-drained and fertile soils in the eastern United States, it grows wild as a deciduous understory tree.

The unripe fruit of this plant contain annonaceous acetogenins, which have anticancer, pesticidal, antimalarial, anthelmintic, piscicidal, antiviral and antibacterial properties, indicating a wide range of possible beneficial uses (Layne 1996). However, the minimal amount of biomass available for extraction has made it difficult to commercially commercialize these chemicals based on extracts.

Beginning in the early 1900s, there was a surge in interest in pawpaw cultivation. For instance, the American Genetic Association presented a $100 prize in 1916, with $40 to $50 going to the largest pawpaw tree and $60 belonging to the tree, no matter how big, that produced the greatest fruit. However, pawpaw is still in its infancy as a commercial fruit crop, despite its great promise throughout the years as a new, high-value specialty crop. Local marketplaces and direct sales to eateries and other upscale niche clients likely offer Pawpaw the biggest market potential at the moment (Pomper et al. 2009; Pomper and Layne 2010).

### *2.5.3* Averrrhoa bilimbi *Linn.*

*A. bilimbi*, often known as Bilimbi, is a member of the Oxalidaceae family. Ibn Rushd, often known as Averroes, was an Arab philosopher, physician and Islamic jurist who inspired the species Averrhoa. The starfruit, or carambola, Averrhoa carambola, is closely related to *A. bilimbi.* It is said to be endemic to the Indonesian Moluccas and West Malaysia, but it originated in Southeast Asia. Malaysia, Indonesia, Singapore, the Philippines, Thailand, Bangladesh, Myanmar and India are among the countries that grow it (Saini 2016). The stem and trunk are where the fruits are produced. The greenish-colored fruits have solid, delicious flesh that softens as it ripens. The fruit juice is very acidic and sour (Kumar et al. 2013).

They can be consumed raw or are frequently used to make pickles and to replace tamarind in recipes. Typically, mature fruits are chopped, sun-dried and stored for later use, or they are preserved in sugar (Bhaskar and Shantaram 2013). The sour fruit juice is utilized to make a beverage. Bilimbi fruits are excellent in removing iron rust from metals and are used to clean utensils. Additionally, they are low in calories, high in antioxidants and rich in vitamin C, all of which are extremely good for human health (Dangat et al. 2014).

### *2.5.4* Prosopis cineraria *(L.) Druce*

*Prosopis cineraria* (L.) Druce, a scientific term for Khejri, is a member of the Leguminosae family and the Mimosoideae subfamily. In addition to being the state tree of Rajasthan, it is also referred to as the Golden Tree of the Desert, the Kalpvriksha of the desert, the King of the Desert, the Love Tree, and the Pride of the Desert. This tree is extremely multifunctional, generating food, fuel, fiber, animal feed, and playing an instrumental role in agroforestry (Panwar et al. 2014). In local communities, it also has great mythical significance. Because of their exceptional environmental adaptability, khejri trees can flourish in a variety of climates.

The khejri tree is indigenous to the gently sloping ravine regions and the northwest Indian plains. About 44 species of plants in the genus *Prosopis* are found mostly in arid regions of Africa, Southwest Asia, and the Americas, which include Patagonia and western North America. In addition to portions of India, namely Rajasthan, Haryana, Punjab, Gujarat, Western Uttar Pradesh, and the drier Deccan, which stretches as far south as Tuticorin, the Prosopis cineraria tree flourishes in the arid and dry regions of Arabia (Rani et al. 2013).

Ripe pods are eaten as fruits, whereas unripe pods are eaten as vegetables. Locally known as Sangri, pods are regarded as a dry fruit of the desert and are essential part of the traditional Rajasthani cuisine panchkuta (Malik et al. 2013). It has 4–9% crude protein, 5–16% sugar, iron, vitamins A and C, and other microminerals in its pods. Sangri pod flour is used with wheat flour to make baked products including bread (chapati). Additionally nutrient-dense, unripe fruit are used to make pickles and curries (Pareek et al. 2015).

## 2.6 Conclusion

Numerous uncommon edible plants which are also referred as neglected crops have medicinal and nutraceutical qualities that make them perfect for commercialization. They are a valuable resource for the transformation of food systems since they can also support healthy, sustainable diets and sustainable agricultural intensification. As a result, they can accomplish several goals. A number of species that are currently endangered or in danger of becoming extinct are threatened by the unsustainable harvesting and use of natural populations. In order to prevent additional population decreases in the wild, sustainable conservation and cultivation methods are desperately needed. Their pharmacological and nutraceutical qualities offer a strong foundation for formulating commercial food and herbal remedies. It is necessary to improve their value chains in order to meet the needs for pharmaceutical and nutraccutical development; yet, their potential is constrained by the limited research and development that has been done thus far. In order to maximize their economic potential while developing new inclusive value chains, RDI should concentrate on every facet of their value chain. To guarantee that these new value chains are inclusive, equitable, and well-supported by the public and commercial sectors, transdisciplinary methods are required.

## References

Adeboye AS, Emmambux NM (2021) Composition, functional and nutritional quality of marama (Tylosema esculentum) storage root. J Food Sci Technol 58(11):4391–4402

Adoukonou-Sagbadja H, Dansi A, Vodouhè R, Akpagana K (2006) Indigenous knowledge and traditional conservation of fonio millet (Digitaria exilis, Digitaria iburua) in Togo. Biodivers Conserv 15(8):2379–2395

Ahmed A, Khalid N, Ahmad A, Abbasi NA, Latif MSZ, Randhawa MA (2014) Phytochemicals and biofunctional properties of buckwheat: a review. J Agric Sci 152(3):349–369

Animasaun DA, Awujoola KF, Oyedeji S, Morakinyo JA, Krishnamurthy R (2018) Diversity level of genomic microsatellite among cultivated genotypes of Digitaria species in Nigeria. Afr Crop Sci J 26(2):305–313

Animasaun DA, Adedibu PA, Olawepo GK, Oyedeji S (2023) Fonio millets: an underutilized crop with potential as a future smart cereal. In: Neglected and underutilized crops. Academic Press, pp 201–219

Aruquipa R, Trigo R, Bosque H, Mercado G, Condori J (2016) El Isaño (Tropaeolum tuberosum) un cultivo de consumo y medicina tradicional en Huatacana para el beneficio de la población boliviana. Revista de Investigación e Innovación Agropecuaria y de Recursos Naturales 3(2):146–151

Augspole I, Duma M, Ozola B, Cinkmanis I (2017) Phenolic profile of fresh and frozen nettle, goutweed, dandelion and chickweed leaves. https://doi.org/10.22616/foodbalt.2017.028

Ballogou VY, Soumanou MM, Toukourou F, Hounhouigan JD (2013) Structure and nutritional composition of fonio (Digitaria exilis) grains: a review. Int Res J Biol Sci 2(1):73–79

Bassey SO, Chinma CE, Ezeocha VC, Adedeji OE, Jolayemi OS, Alozie-Uwa UC, Adie IE, Ofem SI, Adebo JA, Adebo OA (2023) Nutritional and physicochemical changes in two varieties of fonio (Digitaria exilis and Digitaria iburua) during germination. Heliyon 9(6):e17452

Bhaskar B, Shantaram M (2013) Morphological and biochemical characteristics of Averrhoa fruits. Int J Pharma Chem Biol Sci 3(3):924–928

Bordoni A, Danesi F, Di Nunzio M, Taccari A, Valli V (2017) Ancient wheat and health: a legend or the reality? A review on KAMUT khorasan wheat. Int J Food Sci Nutr 68(3):278–286

Brinkkemper O (2015) Smyrnium olusatrum L.(alexanders): an ancient kitchen herb from late medieval Rotterdam (The Netherlands). Veg Hist Archaeobotany 24(1):249–252

Busch JM, Savage GP (2000) Nutritional composition of ulluco (Ullucus tuberosus) tubers. In: Proceedings of the nutrition society of New Zealand, vol 25, pp 55–65

Catarino L, Romeiras MM, Bancessi Q, Duarte D, Faria D, Monteiro F, Moldão M (2019) Edible leafy vegetables from West Africa (Guinea-Bissau): consumption, trade and food potential. Foods 8(10):493

Choquechambi LA, Callisaya IR, Ramos A, Bosque H, Mújica A, Jacobsen SE, Sørensen M, Leidi EO (2019) Assessing the nutritional value of root and tuber crops from Bolivia and Peru. Foods 8(11):526

Chung IM, Paudel N, Kim SH, Yu CY, Ghimire BK (2020) The influence of light wavelength on growth and antioxidant capacity in Pachyrhizus erosus (L.) urban. J Plant Growth Regul 39(1):296–312

Conversa G, Botticella L, Lazzizera C, Bonasia A, Elia A (2024) Ecophysiological and nutritional characterisation of two morphotypes of Cakile maritima subsp. maritima Scop. from Puglia region, Southern Italy. Front Plant Sci 15:1397852

Crozier A, Jaganath IB, Clifford MN (2009) Dietary phenolics: chemistry, bioavailability and effects on health. Nat Prod Rep 26(8):1001–1043

Dangat BT, Shinde AA, Jagtap DN, Desai VR, Shinde PB, Gurav RV (2014) Mineral analysis of Averrhoa bilimbi L. a potential fruit. Asian J Pharm Clin Res 7(3):150–151

Daniela C, Mirko B, Maria PA, Costantina FLM (2010) Morpho-functional adaptations in Cakile maritima Scop. subsp. maritima: comparation of two different morphological types. Caryologia 63(4):411–421

Dębia K, Dzięcioł M, Wróblewska A, Janda-Milczarek K (2025) Goutweed (Aegopodium podagraria L.)—an edible weed with health-promoting properties. Molecules 30(7):1603

Espín S, Villacrés E, Brito Grandes B (2004) Caracterización físico–química, nutricional y funcional de raíces y tubérculos andinos

Flores HE, Walker TS, Guimarães RL, Bais HP, Vivanco JM (2003) Andean root and tuber crops: underground rainbows. HortScience 38(2):161–167

Gandour M, Hessini K, Abdelly C (2008) Understanding the population genetic structure of coastal species (Cakile maritima): seed dispersal and the role of sea currents in determining population structure. Genet Res 90(2):167–178

Gebauer J, El-Siddig K, Ebert G (2002) Baobab (Adansonia digitata L.): a review on a multipurpose tree with promising future in The Sudan. Gartenbauwissenschaft 67(4):155–160

Gunasekera L (2008) Sessile joyweed (Alternanthera sessilis): a popular leafy vegetable in South East Asia but federal noxious weed in USA. In: Proceedings of the 16th Australian weeds conference, Cairns Convention Centre, North Queensland, Australia, pp 347–348
Guo X, Zhu K, Zhang H, Yao H (2010) Anti-tumor activity of a novel protein obtained from tartary buckwheat. Int J Mol Sci 11(12):5201–5211
Halder S, Sharma A (2017) A review on Urtica dioica L. World J Pharm Sci 6:404–421
Hamdi D, Hafidi A, Lemaire JJ, Messaoud C (2025) A comparative study of secondary metabolites profiling and biological activity of Smyrnium olusatrum L. leaf, flower and fruit. Nat Prod Res 39(13):3806–3820
Hammer K, Teklu Y (2008) Plant genetic resources: selected issues from genetic erosion to genetic engineering. J Agric Rural Dev Tropics Subtropics (JARTS) 109(1):15–50
Hamunyela MH, Nepolo E, Emmambux MN (2020) Proximate and starch composition of marama (Tylosema esculentum) storage roots during an annual growth period. S Afr J Sci 116(5–6):1–6
Hartl A, Vogl CR (2002) Dry matter and fiber yields, and the fiber characteristics of five nettle clones (Urtica dioica L.) organically grown in Austria for potential textile use. Am J Altern Agric 17(4):195–200
Heywood VH (1972) The biology and chemistry of the Umbelliferae. Taxon 21:362
Holse M, Husted S, Hansen Å (2010) Chemical composition of marama bean (Tylosema esculentum)—a wild African bean with unexploited potential. J Food Compos Anal 23(6):648–657
Hormaza JI (2014) The pawpaw, a forgotten North American fruit tree. Arnoldia 72(1):13–23
Ingram J, Zurek M (2019) Food systems approaches for the future. In: Agriculture & food systems to 2050: global trends, challenges and opportunities, pp 547–567
Iyaka YA, Idris S, Alawode RA, Bagudo BU (2014) Nutritive values and some mineral elements of Amaranthus hybridus and Adansonia digitata leafy vegetables. J Food Sci 71:25106–25109
Jackson JC, Duodu KG, Holse M, de Faria MDL, Jordaan D, Chingwaru W, Hansen A, Cencic A, Kandawa-Schultz M, Mpotokwane SM, Chimwamurombe P (2010) The morama bean (Tylosema esculentum): a potential crop for southern Africa. Adv Food Nutr Res 61:187–246
Jan KN, Zarafshan K, Singh S (2017) Stinging nettle (Urtica dioica L.): a reservoir of nutrition and bioactive components with great functional potential. J Food Meas Charact 11(2):423–433
Junior AG, de Souza P, dos Reis Livero FA (2019) Plinia cauliflora (Mart.) Kausel. a comprehensive ethnopharmacological review of a genuinely Brazilian species. J Ethnopharmacol 245:112169
Kodahl N (2020) Sacha inchi (Plukenetia volubilis L.)—from lost crop of the Incas to part of the solution to global challenges? Planta 251(4):80
Kodahl N, Sørensen M (2021) Sacha inchi (Plukenetia volubilis L.) is an underutilized crop with a great potential. Agronomy 11(6):1066
Koreissi-Dembélé Y, Fanou-Fogny N, Hulshof PJ, Brouwer ID (2013) Fonio (Digitaria exilis) landraces in Mali: nutrient and phytate content, genetic diversity and effect of processing. J Food Compos Anal 29(2):134–143
Ksouri R, Megdiche W, Debez A, Falleh H, Grignon C, Abdelly C (2007) Salinity effects on polyphenol content and antioxidant activities in leaves of the halophyte Cakile maritima. Plant Physiol Biochem 45(3–4):244–249
Kumar KA, Gousia SK, Anupama MAJNLL, Latha JNL (2013) A review on phytochemical constituents and biological assays of Averrhoa bilimbi. Int J Pharm Pharm Sci Res 3(4):136–139
Kwon RH, Ha BJ (2012) Increased flavonoid compounds from fermented Houttuynia cordata using isolated six of bacillus from traditionally fermented Houttuynia cordata. Toxicol Res 28(2):117–122
Kyrbassova EA, Baitasheva GU, Danilov MP, Dyuskalieva GU, Abdrassulova ZT, Adenova BE, Saparov A (2018) Anatomical-morphological and phytochemical study of medicinal plant Aegopodium Podagraria L. growing in Kazakhstan. Int J Pharm Res 10(4):689–697
Layne DR (1996) The pawpaw [Asimina triloba (L.) Dunal]: a new fruit crop for Kentucky and the United States. HortScience 31(5):777–784

Leidi EO, Altamirano AM, Mercado G, Rodriguez JP, Ramos A, Alandia G, Sørensen M, Jacobsen SE (2018) Andean roots and tubers crops as sources of functional foods. J Funct Foods 51:86–93
Lim TK (2012) Edible medicinal and non-medicinal plants, vol 1. Springer, Dordrecht, pp 285–292
Lim TK (2014) Arracacia xanthorrhiza. In: Edible medicinal and non medicinal plants: volume 9, modified stems, roots, bulbs. Springer Netherlands, Dordrecht, pp 361–366
Lin SC, Lin CC, Lin YH, Shyuu SJ (1994) Hepatoprotective effects of Taiwan folk medicine: Wedelia chinensis on three hepatotoxin-induced hepatotoxicity. Am J Chin Med 22(02):155–168
Loaiza MEE (2013) Manejo Agroecologico del Cultivo de Sacha Inchi (Plukenetia volubilis L.). Universidad Agraria del Ecuador, Guayaquil
Luitel DR, Siwakoti M, Joshi MD, Rangaswami M, Jha PK (2021) Potential suitable habitat of Buckwheat (Fagopyrum spp) under the climate change scenarios in Nepal. J Crop Sci Biotechnol 24(4):401–410
Lukitaningsih E (2014) Bioactive compounds in bengkoang (Pachyrhizus erosus) as antioxidant and tyrosinase inhibiting agents. Indonesian J Pharm 25(2):68
Ma L, Wu F (2003) Advance of chemical constituents and bioactivity of Saururuaceae plants. Zhongguo Zhong Yao Za Zhi:196–198
Malik S, Mann S, Gupta D, Gupta RK (2013) Nutraceutical properties of Prosopis cineraria (L.) druce pods: a component of "Panchkuta". J Pharmacogn Phytochem 2(2)
Meot-Duros L, Le Floch G, Magné C (2008) Radical scavenging, antioxidant and antimicrobial activities of halophytic species. J Ethnopharmacol 116(2):258–262
Messaoud C, Benabdallah A, Boussaid M (2013) Phytochemicals, antioxidant and α-amylase inhibitory activities of Smyrnium olusatrum L. leaf, flower and fruit. Sci Bull:212–220
Mhlaba ZB, Mashilo J, Shimelis H, Assefa AB, Modi AT (2018) Progress in genetic analysis and breeding of tepary bean (Phaseolus acutifolius A. Gray): a review. Sci Hortic 237:112–119
Moghaddam SM, Oladzad A, Koh C, Ramsay L, Hart JP, Mamidi S, Hoopes G, Sreedasyam A, Wiersma A, Zhao D, Grimwood J (2021) The tepary bean genome provides insight into evolution and domestication under heat stress. Nat Commun 12(1):2638
Molberg Ø, Uhlen AK, Jensen T, Flæte NS, Fleckenstein B, Arentz–Hansen H, Raki M, Lundin KE, Sollid LM (2005) Mapping of gluten T-cell epitopes in the bread wheat ancestors: implications for celiac disease. Gastroenterology 128(2):393–401
Nguanchoo V, Srisanga PRACHAYA, Swangpol S, Prathanturarug S, Jenjittikul T (2014) Food plants in Hmong cuisine in Northern Thailand. Thai J Bot 6:131–145
Nyam DD, Kwon-Ndung EH, Wuyep AP (2017) RFLP-based gel electrophoregram of DNA primers for Acha (Fonio) characterisation from Nigeria. Int J Innov Approaches Agric Res 1:51–70
ole Alwis GH, Wijesekera RD, Jayasekera TADNAK (2006) Mukunuwenna, a leafy vegetable grown in Sri Lanka. Bull Environ Contam Toxicol 76:119–125
Omotayo AO, Aremu AO (2021) Marama bean [Tylosema esculentum (Burch.) A. Schreib.]: an indigenous plant with potential for food, nutrition, and economic sustainability. Food Funct 12(6):2389–2403
Orav A, Viitak A, Vaher M (2010) Identification of bioactive compounds in the leaves and stems of Aegopodium podagraria by various analytical techniques. Procedia Chem 2(1):152–160
Padulosi S, Hodgkin T, Williams JT, Haq N (2002) Underutilized crops: trends, challenges and opportunities in the 21st century. In: Managing plant genetic diversity. Proceedings of an international conference, Kuala Lumpur, Malaysia, 12-16 June 2002. CABI Publishing, Wallingford UK, pp 323–338
Padulosi S, Amaya K, Jäger M, Gotor E, Rojas W, Valdivia R (2014) A holistic approach to enhance the use of neglected and underutilized species: the case of Andean grains in Bolivia and Peru. Sustainability 6(3):1283–1312
Panwar D, Pareek K, Bharti CS (2014) Unripe pods of Prosopis cineraria used as a vegetable (sangri) in Shekhawati region. Int J Sci Eng Res 5(2):892–895
Pareek AK, Garg S, Kumar M, Yadav SM (2015) Prosopis cineraria: a gift of nature for pharmacy. Int J Pharm Sci Res 6(6):958–964

Patel N, Krishnamurthy R (2021) In-vitro phytochemical screening and bioactivity of Moringa oleifera accessions. Biosci Biotechnol Res Commun 14(1):335–339

Patel N, Krishnamurthy R (2023) Moringa oleifera accessions: perspectives and application as nutraceuticals and phytomedicines. In: Bioprospecting of tropical medicinal plants. Springer Nature Switzerland, Cham, pp 463–479

Patel N, Tarjani C, Ramar K (2023) Phytochemical constituents and antibacterial activity of different plant parts extracts of wild Moringa oleifera Lam. Ann Biol 39(2):311–314

Patel N, Oyedeji S, Animasaun DA, Olorunmaiye KS, Krishnamurthy R (2025) The potential of underutilized and undervalued Moringa Oleifera plant accessions for nutritional and food security. In: Food security and nutrition. CRC Press, pp 85–96

Petrelli R, Ranjbarian F, Dall'Acqua S, Papa F, Iannarelli R, Kamte SLN, Vittori S, Benelli G, Maggi F, Hofer A, Cappellacci L (2017) An overlooked horticultural crop, Smyrnium olusatrum, as a potential source of compounds effective against African trypanosomiasis. Parasitol Int 66(2):146–151

Piergiovanni AR, Simeone R, Pasqualone A (2009) Oriental wheat, an underutilised tetraploid wheat species. A case study: nutritional and technological traits of Kamut. Food 3(1):33–38

Pomper KW, Layne DR (2010) The North American pawpaw: botany and horticulture. Hortic Rev 31:349–382

Pomper KW, Lowe JD, Crabtree SB, Keller W (2009) Identification of annonaceous acetogenins in the ripe fruit of the North American pawpaw (Asimina triloba). J Agric Food Chem 57(18):8339–8343

Porch TG, Cichy K, Wang W, Brick M, Beaver JS, Santana-Morant D, Grusak MA (2017) Nutritional composition and cooking characteristics of tepary bean (Phaseolus acutifolius Gray) in comparison with common bean (Phaseolus vulgaris L.). Genet Resour Crop Evol 64(5):935–953

Pradhan S, Rituparna S, Dehury H, Dhall M, Singh YD (2023) Nutritional profile and pharmacological aspect of Houttuynia cordata Thunb. and their therapeutic applications. Pharmacol Res-Modern Chinese Med 9:100311

Randall RE (2003) Smyrnium olusatrum L. J Ecol 91(2):325–340

Rani B, Singh U, Sharma R, Gupta A, Dhawan NG, Sharma AK, Sharma S, Maheshwari RK (2013) Prosopis cineraria (L.) Druce: a desert tree to brace livelihood in Rajasthan. Asian J Pharm Healthc Res 5(2):58–64

Rao I, Beebe S, Polania J, Ricaurte J, Cajiao C, Garcia R, Rivera M (2013) Can tepary bean be a model for improvement of drought resistance in common bean? Afr Crop Sci J 21(4)

Ravi V, Raju S, Murugesan P (2025) Mexican yam bean (Pachyrhizus erosus L. Urban): a neglected leguminous root crop with emerging potential for food security and industrial applications. Crop Sci 65(3):e70057

Roca WM, Ynouye C, Manrique I, Arbizu C, Gomez R (2007) Indigenous andean root and tuber crops: new foods for the new millennium. Chron Hortic 47(4):13–19

Rodzi NARM, Lee LK (2022) Sacha Inchi (Plukenetia Volubilis L.): recent insight on phytochemistry, pharmacology, organoleptic, safety and toxicity perspectives. Heliyon 8(9)

Rutto LK, Xu Y, Ramirez E, Brandt M (2013) Mineral properties and dietary value of raw and processed stinging nettle (Urtica dioica L.). Int J Food Sci 2013(1):857120

Saini S (2016) A review on phytochemistry and pharmacology of averrhoa bilimbi linn. Int Educ Rese J 2(1):71–76

Savinov AB, Erofeeva EA, Nikitin YD (2019) Morphological variability and biochemical parameters of leaves in cenopopulations of Aegopodium podagraria L.(Apiaceae, Apiales) under various levels of soil contamination by heavy metals. Biol Bull 46(10):1347–1352

Sayed-Ahmad B, Talou T, Saad Z, Hijazi A, Merah O (2017) The Apiaceae: ethnomedicinal family as source for industrial uses. Ind Crops Prod 109:661–671

Schinkel C, Gepts P (1988) Phaseolin diversity in the tepary bean, Phaseolus acutifolius A. Gray. Plant Breed 101(4):292–301

Shaheen N, Nawaz S, Nazir A (2025) Common buckwheat: a neglected plant with great health benefits. Innov STEAM: Res Educ 3(1)
Silva PHAD, Faria FCD, Tonon B, Mota SJD, Pinto VT (2008) Evaluation of the chemical compostion of wine produced from jabuticaba (Myrciaria jabuticaba). Quim Nova 31:595–600
Taylor K (2009) Biological flora of the British Isles: Urtica dioica L. J Ecol 97(6):1436–1458
Thornhill SM, Kelly AM (2000) Natural treatment of perennial allergic rhinitis. Altern Med Rev 5(5):448–454
Ulian T, Flores C, Lira R, Mamatsharaga A, Mogotsi KK, Muthoka P, Ngwako S, Nyamongo DO, Omondi W, Sanogo AK, Sanogo S (2019) Wild plants for a sustainable future. Kew Publisher, p 486
Valyova M, Tashev A, Stoyanov S, Yordanova S, Ganeva Y (2016) In vitro free-radical scavenging activity of Aegopodium podagraria L. and Orlaya grandifl ora (L.) Hoffm.(Apiaceae). J Chem Technol Metall 51:271–274
Villa TCC, Maxted N, Scholten M, Ford-Lloyd B (2005) Defining and identifying crop landraces. Plant Genet Resour 3(3):373–384
Wang ZH, Gao L, Li YY, Zhang Z, Yuan JM, Wang HW, Zhang L, Zhu L (2007) Induction of apoptosis by buckwheat trypsin inhibitor in chronic myeloid leukemia K562 cells. Biol Pharm Bull 30(4):783–786
Wang S, Zhu F, Kakuda Y (2018) Sacha inchi (Plukenetia volubilis L.): nutritional composition, biological activity, and uses. Food Chem 265:316–328
Wiwart M, Suchowilska E, Kandler W, Sulyok M, Groenwald P, Krska R (2013) Can polish wheat (Triticum polonicum L.) be an interesting gene source for breeding wheat cultivars with increased resistance to Fusarium head blight? Genet Resour Crop Evol 60(8):2359–2373
Wu SB, Long C, Kennelly EJ (2013) Phytochemistry and health benefits of Jaboticaba, an emerging fruit crop from Brazil. Food Res Int 54(1):148–159
Yu X, Gao Y, Xu Y, Guo X, Guo L, Tan T, Liu F, Wan Y (2022) Study of the contents of analogues of aristolochic acid in Houttuynia cordata by ultra-high performance liquid chromatography tandem mass spectrometry. Foods 11(3):302
Zhou ML, Bai DQ, Tang Y, Zhu XM, Shao JR (2012) Genetic diversity of four new species related to southwestern Sichuan buckwheats as revealed by karyotype, ISSR and allozyme characterization. Plant Syst Evol 298(4):751–759
Zhu F, Cui R (2020) Comparison of physicochemical properties of oca (Oxalis tuberosa), potato, and maize starches. Int J Biol Macromol 148:601–607

# Chapter 3
# The Intersection of Health and Sustainability

**Abstract** There are negative effects on the environment and human health as a result of the increased demand for food from a growing population. The growing popularity of sustainable diets and eating could open up new chances for nutritionists to gain expertise. There are connections between how eating habits can support environmental and health objectives that are being discovered by ongoing research. However, this field presents a challenge for the nutrition professions because pertinent education frequently draws from a less recognizable body of research than traditional nutritional sciences and requires politically motivated, organizational oriented thinking. In order to promote the prospective significance of nutrition professionals, new evidence evaluations and instructional input are crucial as impetus to meet sustainability agenda goals increases. In order to attain sustainability literacy for nutrition professionals, this chapter evaluates sustainable eating and plant diets and takes into account the importance of improved instruction that is more regularly incorporated into nutrition curriculum.

**Keywords** Chronic Diseases · Eating habits · Edible Plants · Nutrition · Sustainability

## 3.1 Introduction

Plant-based diets are suggested for environmental sustainability and are linked to excellent health. The acronym plant-based diet refers to a broad range of dietary regimens that prioritize plants. Foods such fruits, vegetables, wholegrains, legumes, nuts and seeds as well as plant based substitutes; minimize or eliminate products originating from animals. Although figures from national food consumption surveys continue to show that more than 90% of individuals in all population groups still consume meat, news outlets, consumer organizations and vegan and meat-free societies in the Western world are indicating an upsurge toward the growth of Plant based consumers (Arshad et al. 2020). The number of plant based diet users is often unclear and challenging to determine because of the wide range of definitions of

 
N. Patel et al., *The Future of Food: Exploring Edible Plants for a Sustainable Diet*,
https://doi.org/10.1007/978-3-032-15027-1_3

diets based on plants, even though these numbers might not be representative of other diets that contain minor amounts of meat and/or dairy (Sanders 1999).

Traditionally, malnutrition in underdeveloped nations has been linked to a restricted diet and an excessive reliance on one primary meal that is deficient in nutrients. There is little influence over what people in poor nations eat. However, a growing number of individuals are choosing vegetarianism over meat in affluent nations like the UK. Eating meat has consequences for the environment, welfare, and economy in addition to being a moral and personal choice (Williams and Patel 2017). Numerous health arguments have been made in favor of consuming less meat overall, including the links between excessive meat and animal fat intake and the risks of heart disease and cancer.

It's crucial to keep in consideration that meat can spread food-borne illnesses while thinking about the risks that food poses to an individual's health. In developed nations, the majority of food-borne illnesses are spread by meat, eggs, seafood, and milk. The only way to bring worldwide consumption of meat closer to the level witnessed in rich nations now would presumably need to employ intensive animal production techniques, which are inherently riskier. The current dominance of food distribution and processing raises the risk of zoonotic disease transmission, as seen in the bovine spongiform encephalopathy outbreak (Ewy et al. 2022).

To cope up with this a substantial contribution is favoured by uncommon/wild edible plants, which are a rich source of beneficial nutrients. The potential of these plants to prevent malnutrition, increase nutritional diversity and advance health is highlighted by the growing global recognition of their inherent value, which is impacted by religious beliefs and dietary preferences. In addition to adding nutritious value to the world's food supply these uncommon edible plants encourage environmentally friendly behaviour and ease the burden on overgrazed areas (Ostfeld 2017). These plants are ingrained in many cultures as traditional food staples that fit in with regional eating preferences and views. They combat hunger by providing vital micronutrients that are lacking in processed diets and are resilient to climate change. The secret to improving global food security and maintaining health is to recognize and promote their nutritional benefits in line with local preferences and sustainable practices.

## 3.2 The Nutritional Benefits of Plant Based Diets

The environment and nutrition are undoubtedly being harmed by the mainstream agricultural and food production systems. A significant obstacle confronting humanity at the moment is ensuring that everyone has access to enough food that is sustainable, nourishing, healthful, and reasonably priced (Caron et al. 2018). Even if the percentage of the world's population that goes hungry has decreased by half since the early Green Revolution, trends in food insecurity and hunger constantly show that populations are still undernourished (Willett et al. 2019).

Failure to implement focused mitigation tactics in the farming industry will ultimately lead to a 55–85% spike in overall ecological stresses and a breakdown of important ecosystem processes, according to a recent review of the environmental impacts of food systems by Springmann et al. (2018). In contrast to adopting plant-based diets, which raised these percentages to 50% and 5–20%, respectively, the investigation found that switching to better diets solely might cut greenhouse gas emissions and other environmental consequences by 28% respectively. These numbers reinforce our belief that a shift in the food system is necessary to go beyond increasing food production and rather seek for methods to nurture individuals in methods that can be perpetuated environmentally, financially, and ethically (Fig. 3.1) (Haddad et al. 2016).

In addition to providing important macronutrients like proteins, lipids, and carbs uncommon and underused edible plants also offer micronutrients like vitamins and minerals, as well as functional non-nutrients that support nutritional health. For many nations, the variety of foods is a natural resource, but most, if not all, do not make sufficient use of it. Both at the species level and at the level of variety within a species, these plants can contribute to varied and healthful meals. Numerous researchers studying food composition are aware that variations in nutrient content within the same species might often differ more than those between species (Romanelli et al. 2015). Regretfully, this varietal level has not been thoroughly investigated in the majority of studies. The majority of food composition data have been restricted to an aggregate level, frequently neglecting the substantial compositional differences related to agroecological zone, seasonality, and, most importantly, genetic diversity. The nutritional science usually just takes inter-species diversity into account when analysing diet diversity (WHO 2015).

Traditional variants of locally significant species frequently have a greater nutritious profile than more widely consumed cultivars that predominate the market, according to research from the Pacific. However, for various socioeconomic or political reasons, many of these traditional, nutrient-rich varieties have been neglected. Certain indigenous banana varieties have pro-vitamin A carotenoid

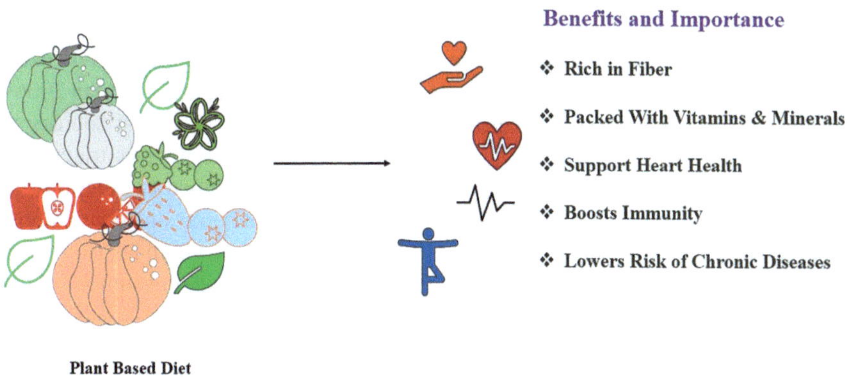


**Fig. 3.1** Benefits of Plant Based Diet

contents that were more than 1000 times higher than those of the common Cavendish variety (Fanzo et al. 2013). Researchers have documented studies on uncommon wild legume plant species worldwide, including North America, Africa, Korea, China, India, Mexico and Nigeria. Legumes are one of the most important and valuable plant sources for human food. The nutritional availability of more than 15 underused legumes in Africa were examined in a study, and the findings showed that because they are rich in proteins, vital amino acids, minerals, and energy, they are employed as sources for biofortification (Bhat and Karim 2009). Additionally, these legumes were screened for lauric, myristic, palmitic, arachidic, lignoceric, myristoleic, elaidic, and heneicosanoic acids. Indians frequently eat the weed legume, *Indigofera glandulosa*, which was discovered in North Australia, Indonesia, and India. Experts have recently found that this species is rich in proteins, vital amino acids, Vitamin B, and high caloric value (Ghane et al. 2010).

It is believed that at least 900 to 1000 million individuals use wild or uncommon vegetation, which are essential to many rural sustenance households. For instance, in Ghana, people eat the leaves of more than 90 uncommon edible plant species along with fruits of another 190 to 200 species. In rural Swaziland, over 200 kinds of wild plants make up a larger portion of the food than domesticated varieties (Deokule and Aberoumand 2009). In Thailand, Malaysia, and India, some 100 varieties of wild plants have been recognized as urgent nourishment sources. Food plants found in forests have been staples in many Indian states (Bacchetta et al. 2016).

Numerous bioactive compounds with medicinal and antioxidant qualities are also abundant in wild plants. The primary and significant risk factor for heightened susceptibility to infectious diseases and a major risk factor for a number of non-communicable diseases is poor nutrition. Wild edible species are used as food, vegetables, beverages, cooking oils (made from seeds) sauces and spices, medicine and industrial products in indigenous tribal communities. The cultural, social, and religious belief systems of rural communities would be incomplete without these wild food plants. These wild food plants could be domesticated to increase both their use and conservation (Singh et al. 2016; Joshi et al. 2018).

Investigations have indicated that "nutraceuticals," which denote both nutritional and medicinal qualities, originate from edible wild plants. In addition to being cholesterol-free and comparatively low in calories, the wild edible plants have a distinct flavor and taste. Plants make up a sizable portion of the normal dietary component, which is essential to a healthy diet worldwide (Flyman and Afolayan 2006). Green leafy vegetables, both traditional and non-traditional, are great sources of minerals that, when consumed frequently, can help people with micronutrient deficiencies at the lowest possible cost. Furthermore, when consumed in excess, these veggies have a number of toxicological effects on the human body and contain both essential and harmful metals in varying amounts. According to nomads from rural Africa, 60–65 wild foods derived from 50 to 53 plant species are consumed locally and gender and ethnic group differences exist in the understanding of wild foods (Satter et al. 2016). The majority of the time, wild plant foods (fruits, vegetables, and spices) are accessible in Cameroon's cities, and people spend a significant portion of their dietary expense on them. The majority of the 105 species discovered in an exploratory study on wild food plants conducted in Uganda are used locally, however some are traded (Karasawa and Mohan 2018).

## 3.3 Eco Geographical Diversity and Contributions to Human Nutrition

National contributions to the recent State of the World's Biodiversity for Food and Agriculture of the Food and Agriculture Organization of the United Nations confirms the widespread use of uncommon edible plants in multiple nations. Out of the 91 countries that provided data for the report, 69 countries identified a total of 50 to 55 wild plant species that improve nutrition and food security in their respective nations while also enhancing the diversity and health of diets (Penafiel et al. 2011). However, the true contribution of these ecosystems to national economies in several parts of the planet is also generally neglected in national statistics, and the number could be significantly higher (Penafiel et al. 2019).

The traditional diets of indigenous populations residing in forest areas of Ecuador, one of the nations with the seven most diverse countries, heavily rely on wild edible fruits and plants that are gathered from a variety of environments. According to the reports collected from the entire country, the Andean Kichwa moms of Arosemena Tola (Napo) and the Andean indigenous tribes of Guasaganda (Cotopaxi) have a culinary reliance on 10 and 49 WPFs, respectively (Jacob et al. 2020). The majority of this diversity is of local worth, and its potential nutritional and economic value is still untapped. However, some species, including *Bertholletia excelsa* and *Euterpe oleracea*, are of national and geographical importance (Penafiel et al. 2019).

Numerous wild species with nutritional value and commercial potential have been found by the Ministry of the Environment's Plants for the Future Initiative, a prioritizing exercise that aimed to investigate the wealth of Brazil's plant life. Of the 78 native, underrated edible plant species found in the country's five ecoregions, 49 are found only in the wild (mainly fruits and nuts). These species, which are primarily found in forest regions, are either collected from the wild by local communities using customary methods or managed by family farmers (Baldermann et al. 2016). Human well being in the Indo-Pacific is significantly improved by the eating of uncommon wild edible plants. About half of the rural dwellers, especially those in high-altitude communities, consume the kernel of the *Pandanus brosimos*, which is unique to New Guinea (Novaczek 2001; Bourke and Allen 2009). In order to gather the fruit and collect the nuts, the locals move to higher elevations when the fruit reaches maturity. The conventional processing of *Tacca* spp. is performed in Niue, Polynesia. A specialty of the area, starch extracted from the rhizome is used to produce breads and puddings. Uncommon marine seaweeds like *Caulerpa racemosa*, and other edible seaweeds that are still commonly eaten are among them. In the mosaic tropical region of West Sumatra, Indonesia, which is mostly composed of rice fields, residential gardens, cocoa agroforestry and woodlands, the Food, Agrobiodiversity and Diet project has found 80 to 85 wild and uncommon edible plants with the help of indigenous populations (Pawera et al. 2018, 2020).

Under the Royal Botanic Gardens' (UK) MGU Useful Plants Project, local populations in South Africa, Botswana, Mali, Kenya and Mexico collaborated with institutional partners to identify more than 600 edible plant species in the five nations. *Schinziophyton rautanenii*, *Adansonia digitata* and *Tylosema esculentum* were among the species found in Africa (Ulian et al. 2017). There are more than 200 leafy vegetables in Kenya alone, according to investigations conducted by Bioversity International in the early 1990s. These are partially domesticated or native plants that are mostly cultivated for domestic use or private trade, but have experienced a resurgence, especially in peri-urban and urban settings. Depending on the region, indigenous communities in Western Kenya continue to consume African green vegetables. Under Leafy Vegetables program, eleven species were picked for additional study, including *Solanum* spp., *Cleome gynandra*, and *Amaranthus* spp. (Ulian et al. 2019). Wild plant species, particularly vegetables, can significantly lower the expense of nutritious food female and kids aged 5 to 22 months in modelled affordable healthy diets, in addition to bridging the nutrient gap, according to an expense of diet study conducted in Eastern Baringo, Kenya (Sarfo et al. 2020).

Uncommon edible plants are still frequently used in conventional Mediterranean recipes and are utilized by many locals. About 2000 to 2300 distinct plant and fungal species were found in this region alone, of which 900 to 1200 were exclusively utilized locally, according to their compilation of collected Mediterranean edible plants (Rivera et al. 2006). More than 40 wild edible plants were given priority for additional study as part of the Biodiversity project in Turkey. However, in rural Morocco, wherein nearly eighty percent of households reported regularly ingesting wild leafy vegetables, they represent a seasonally significant element of Moroccan diets. *Malva* spp., *Portulaca oleracea*, *Chenopodium* spp., Rumex spp., *Scolymus hispanicus*, and *Nasturtium officinale* are frequently eaten. Many of these species are also used in Turkey. They are often consumed in reasonable portions as a homemade salad or accompanying dish (Hadjichambis et al. 2008; Nassif and Tanji 2013).

## 3.4 Obstacles to the Broader Utilization of Uncommon Edible Plant Species

The underutilization of uncommon edible plants for food and nutritional security can be partially ascribed to a number of economic and non-commercial obstacles that restrict their unrealized potential, as well as a lack of proof and understanding among policymakers as well as stakeholders regarding the significance of wild foods for dietary habits, economic growth, and food security. Insufficient or brief study and outreach funding to assist the development of unconventional, traditional, and indigenous food resources is another factor contributing to their inadequate awareness (Fig. 3.2).

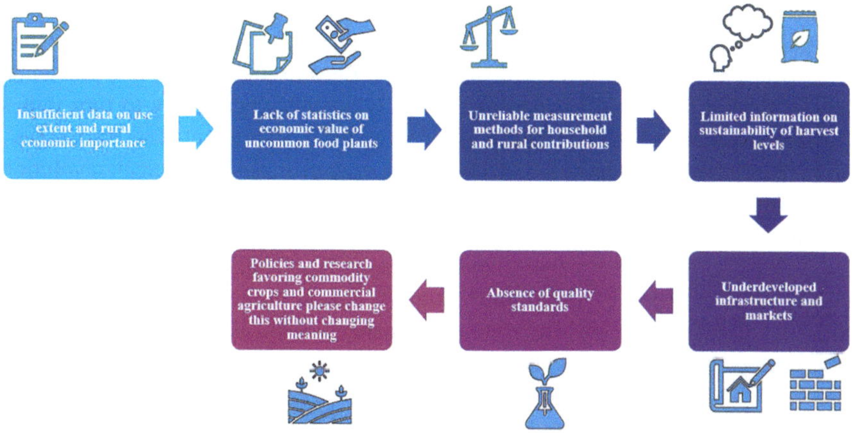


**Fig. 3.2** Obstacles in Utilization of Uncommon Edible Plant Species

## 3.5 Sustainable Eating Patterns

The present food framework is deviated by disparities of accessibility and cannot be sustained. Although there is more than enough food produced to feed the world's 7 billion people, half of them suffer from malnutrition, and almost 2 billion lack essential micronutrients. A global epidemic of illness that costs the £6 billion annually in the United Kingdom is linked to the ensuing dual impact of ailments (obesity and malnutrition), which is partly caused by "the nutritional shift" resulting in higher dietary intakes of sugary foods, trans fats, oils, and packaged meats (Lang 2014). Current eating patterns are expected to significantly harm the environment (such as by reducing biodiversity and increasing pollution) and increasing illness by 2050 if left unregulated (Pretty et al. 2005).

Thus, a major worldwide challenge is putting measures in place to deal with the closely related "nutrition, environment, and health" conflict. Unsustainable climate change is blatantly caused by industrial and animal agriculture. The government has also acknowledged the need to lessen the impact of the strain on water supplies for agricultural usage. There are several ways in which optimizing land usage for food production for example, by lowering the quantity of land needed for meat rearing would benefit the environment. Agriculture and food production practices, as well as dietary intake habits, must alter to meet climate change-related goals and lessen its severe effects (Tilman and Clark 2014).

It is well known that "sustainable eating" is a wide-ranging and intricate topic that sits at the nexus of environmental health, wellness and the culinary industry. The numerous major companies in the food sector have varying interests, and the market is enormous. Improved sector-wide leadership is being called for in order to promote an ecologically sustainable strategy for the food industry (Tansey 2012).

There is broad scientific agreement that human-induced greenhouse gas emissions have an impact on climate change and global warming. In an effort to mitigate the adverse effects of environmental change, 196 parties have pledged to hold global warming to 1.5 °C and restrict the rise in the average world temperature to less than 2 °C over pre-industrial levels (Stocker 2014). Because the coronavirus pandemic has reoriented the world's economies, there is another opportunity to turn things greener and increase the possibility of reaching the 1.5 °C target. The British government, for instance, has outlined a ten-point strategy to achieve a green revolution in industries, which calls for reforming the energy sector, banning the sale of gasoline and diesel vehicles, decarbonizing public transportation, creating greener buildings, investing in carbon capture and storage, and safeguarding the environment (Willett et al. 2019). It is concerning that they left agriculture out of their strategies. Since the primary source of environmental change worldwide is food production, it is arguable that transforming agricultural systems ought to be a major priority (Schwab and Malleret 2020).

Although they occupy about 40% of the planet's area, current agricultural practices are responsible for more than 60% of freshwater use and up to 20% of the total human GHG emissions. As a result, innovations in the agriculture industry has an opportunity to produce significant improvements in sustainability. Adopting a switch to a plant derived food is one potential mode which is gaining more and more attention (Vermeulen et al. 2012). Compared to animal-based diets, plant derived foods have a substantially smaller environmental impact. Any diet regimen that emphasizes the consumption of meal originated from plants and restricts or eliminates the consumption of the majority of animal products is referred to as plant derived foods. Depending on their makeup, these foods can be either good or unhealthy (Steffen et al. 2015). While harmful foods include large amounts of modified and highly altered plant foods like sugar-sweetened beverages, refined grains, sweets and desserts, healthy plant derived food emphasize unprocessed/uncommon edible plant foods like fruits, vegetables, legumes, grains and seeds (Foley et al. 2005).

Food-associated disease is all-time high on a global scale. Over two billion adults are overweight or obese globally. Numerous ongoing medical conditions, such as type 2 diabetes, hypertension, cardiovascular disease, and certain types of cancer, are linked to being overweight or obese. When taken as a whole, these ailments cost society a great deal in terms of lost lives and medical expenses (Zhou et al. 2016). According to the Global Burden Disease (GBD) study, by improving the risk factors for chronic diseases, eating more whole grains, vegetables, nuts, seeds and fruit might avert 1.7 to 4.9 million premature deaths annually. In the span of 25–30 years, type 2 diabetes has almost doubled in frequency worldwide. Diabetes was the cause of $900 billion in medical expenses and six million fatalities in 2021. According to large comprehensive investigations, people who follow plant diet have a much lower incidence of diabetes type 2 (Sun et al. 2022). This trended similarly to that of body mass index, with vegans having the lowest prevalence and non-vegetarians having the highest. The frequency of this disease was intermediate among vegetarians, pescatarians and semi-vegetarians (Tonstad et al. 2009).

When comparing diabetes type 2 risk to plant diet adherence, a meta-analysis published in 2019 of nine potential studies with a total of 300,000 participants revealed a significant inverse relationship. There exist indications that these diets may be a useful treatment for diabetes type 2 in addition to minimizing the condition (Qian et al. 2019). Consuming plant diets was linked to a substantial decrease in hemoglobin A1c when compared to omnivorous control diets, according to a meta analysis of six controlled trials. This hypoglycemic effect is around half of what is seen when metformin, the first-line drug, is prescribed (Yokoyama et al. 2014).

The use of plant diet to avoid overweight and disorders associated with obesity is supported by a substantial number of research and clinical trials. Body mass index had a significant association with the amount of meat and dairy products consumed, according to observational data from the Adventist Health Study-2, which included 41,000 participants (Tonstad et al. 2013). Those who were not vegetarians experienced the most obese Body mass index, followed by semi-vegetarians, pescatarians, vegetarians and vegans. Furthermore, results from the EPIC-Oxford cohort, which included 21,900 men and women, have demonstrated that vegetarians and vegans lose more weight per year than those who eat meat (Turner-McGrievy et al. 2017). The use of PBDs for weight loss is supported by strong data from clinical trials. A systematic review of 15 clinical studies using vegan and vegetarian therapies lasting 4 weeks or longer without caloric limitations was conducted in 2015 by Barnard et al.

Even after adjusting for body mass index, vegans in the Adventist Health Study-2 group had around half of the likelihood of having hypertension compared to omnivores. Despite the presence of certain products derived from animals, a 2021 systematic review comprising 40 controlled studies and 8000 people discovered that plant diets significantly reduce blood pressure at all levels (Pettersen et al. 2012). According to a 2017 comprehensive review of 19 trials with 1400 people, vegetarian diets were substantially linked to lower levels of total cholesterol and low-density lipoprotein cholesterol when compared to omnivorous diet intake (Gibbs et al. 2021).

Worldwide, cardiovascular diseases are the primary cause of death. Around the world, this disease caused estimated 18.6 million in 2019. The consumption of plant diets for preventing cardiovascular diseases is supported by a variety of studies (Roth et al. 2020). According to a 2021 systematic review of prospective cohort investigations with maximum participants, plant diets were linked to 11% lower risk of ischemic heart disease and a 15% lower risk of cardiovascular diseases. According to a different 2021 comprehensive review of prospective cohort studies with 410,000 participants, these diets were linked to an 8% reduced risk of cardiac death and a 10% lower risk of cardiovascular diseases recurrence (Quek et al. 2021).

## 3.6 The True Acceptability of Plant Based Eating

Food acceptability is influenced by interpersonal relationships, accessibility, affordability, sensory characteristics, cultural context and surrounding variables. Although people in advanced nations frequently regard animal products as an essential component of meals and perceive vegetarian and vegan diets as unpleasant, challenging to prepare, and less enjoyable than omnivorous diets, plant diets are typically embraced in low- and low-middle-income countries for social, economic, moral, and/or spiritual reasons (Moreno et al. 2022). However, trends are shifting; in nations with higher incomes, the number of vegans and vegetarians continues to rise, while in developing and low-middle-income countries, like China and India, a shift in diets is occurring as a result of rising poultry and animal product consumption brought on by annual income growth (Popkin et al. 2020).

According to optimization studies, it is possible to strike a balance among cost, dietary adequacy, and ecology, along with some consideration for diet feasibility as previously stated. However, in several instances, the optimal diets deviated significantly from the consumers' current eating patterns; as a result, they might not be embraced and adhered to (Fanzo et al. 2022). For instance, the Galactic Diet calls for a sharp decrease in the intake of meat, eggs, potatoes, and processed foods, while a considerable increase in the consumption of plant-based oils, legumes, and whole grains relative to the global average (Onwezen et al. 2021).

Despite the diversity of eating habits, culinary cultures, and ecological problems, context-specific solutions are needed to promote an ecologically sound plant-based dietary strategy. Since recognition is one of the elements that influences consumer acceptance, a focus on conventional, locally grown foods may increase acceptability while also having beneficial environmental consequences (Martínez-González et al. 2017). For example, the Mediterranean diet model aligns with the cultural diversity associated with the region, the fellowship, and the way of life that is transformed into an authentic act of connection and exchange through food, in addition to being a plant-based dietary pattern correlated with favorable health results. Non-Mediterranean communities can also benefit from the Mediterranean diet's health benefits, which include foods that are readily accessible, naturally grown, and socioeconomically acceptable while sharing a similar nutritive profile. Regarding the Planetary Diet study, the authors highlighted how plant-based dietary habits must be implemented while taking into account local communities' cultural preferences, nutritional availability, and economics (Woodside et al. 2022).

## 3.7 Conclusion

Plant centered diets may be economical, environmentally friendly, socially acceptable, safe and nutrient-dense. However, obstacles must be considered and addressed in order to guarantee this sustainable shift from the current lifestyles to sustainable plant-centered dietary patterns. The construction and encouragement of urban gardens and farmlands, the modernization and efficient distribution of food based guidelines, value addition, development of plant-based substitutes based on regional cuisines and raising the price of unhealthy foods are a few examples of significant actions and initiatives. To address the challenges confronting the world and human health, future inclusive discussions with researchers, legislators, and other stakeholders, including all those involved in food systems, must take place.

## References


Arshad F, Waheed M, Iqbal M, Fatima K, Fatima K (2020) Ethnobotanical assessment of woody flora of district Kasur (Punjab), Pakistan. Ethnobot Res Appl 20:1–13

Bacchetta L, Visioli F, Cappelli G, Caruso E, Martin G, Nemeth E, Bacchetta G, Bedini G, Wezel A, van Asseldonk T, van Raamsdonk L (2016) A manifesto for the valorization of wild edible plants. J Ethnopharmacol 191:180–187

Baldermann S, Blagojević L, Frede K, Klopsch R, Neugart S, Neumann A, Ngwene B, Norkeweit J, Schröter D, Schröter A, Schweigert FJ (2016) Are neglected plants the food for the future? Crit Rev Plant Sci 35(2):106–119

Barnard ND, Levin SM, Yokoyama Y (2015) A systematic review and meta-analysis of changes in body weight in clinical trials of vegetarian diets. J Acad Nutr Diet 115(6):954–969

Bhat R, Karim AA (2009) Exploring the nutritional potential of wild and underutilized legumes. Compr Rev Food Sci Food Saf 8(4):305–331

Bourke MR, Allen B (2009) Nuts. In: Bourke M, Harwood T (eds) Food and agriculture in Papua New Guinea. ANU Press, pp 215–222

Caron P, de Ferrero y Loma-Osorio G, Nabarro D, Hainzelin E, Guillou M, Andersen I, Arnold T, Astralaga M, Beukeboom M, Bickersteth S, Bwalya M (2018) Food systems for sustainable development: proposals for a profound four-part transformation. Agron Sustain Dev 38(4):41

Deokule SS, Aberoumand A (2009) Studies on nutritional values of some wild edible plants from Iran and India. Pak J Nutr 8(1):26–31

Ewy MW, Patel A, Abdelmagid MG, Mohamed Elfadil O, Bonnes SL, Salonen BR, Hurt RT, Mundi MS (2022) Plant-based diet: is it as good as an animal-based diet when it comes to protein? Curr Nutr Rep 11(2):337–346

Fanzo J, Hunter D, Borelli T, Mattei F (2013) Diversifying food and diets. Routledge, London

Fanzo J, Rudie C, Sigman I, Grinspoon S, Benton TG, Brown ME, Covic N, Fitch K, Golden CD, Grace D, Hivert MF (2022) Sustainable food systems and nutrition in the 21st century: a report from the 22nd annual Harvard Nutrition Obesity Symposium. Am J Clin Nutr 115(1):18–33

Flyman MV, Afolayan AJ (2006) The suitability of wild vegetables for alleviating human dietary deficiencies. S Afr J Bot 72(4):492–497

Foley JA, DeFries R, Asner GP, Barford C, Bonan G, Carpenter SR, Chapin FS, Coe MT, Daily GC, Gibbs HK, Helkowski JH (2005) Global consequences of land use. Science 309(5734):570–574

Ghane SG, Lokhande VH, Ahire ML, Nikam TD (2010) *Indigofera glandulosa* Wendl.(Barbada) a potential source of nutritious food: underutilized and neglected legume in India. Genet Resour Crop Evol 57(1):147–153

Gibbs J, Gaskin E, Ji C, Miller MA, Cappuccio FP (2021) The effect of plant-based dietary patterns on blood pressure: a systematic review and meta-analysis of controlled intervention trials. J Hypertens 39(1):23–37

Haddad L, Hawkes C, Webb P, Thomas S, Beddington J, Waage J, Flynn D (2016) A new global research agenda for food. Nature 540(7631):30–32

Hadjichambis AC, Paraskeva-Hadjichambi D, Della A, Elena Giusti M, De Pasquale C, Lenzarini C, Censorii E, Reyes Gonzales-Tejero M, Patricia Sanchez-Rojas C, Ramiro-Gutierrez JM, Skoula M (2008) Wild and semi-domesticated food plant consumption in seven circum-Mediterranean areas. Int J Food Sci Nutr 59(5):383–414

Jacob MCM, Araujo de Medeiros MF, Albuquerque UP (2020) Biodiverse food plants in the semi-arid region of Brazil have unknown potential: a systematic review. PLoS One 15(5):e0230936

Joshi SK, Ballabh B, Negi PS, Dwivedi SK (2018) Diversity, distribution, use pattern and evaluation of wild edible plants of Uttarakhand, India. Defence Life Sci J 3(2):126–135

Karasawa MMG, Mohan C (2018) Fruits as prospective reserves of bioactive compounds: a review. Nat Prod Bioprospect 8(5):335–346

Lang TIM (2014) Sustainable diets: hairshirts or a better food future? Development 57(2):240–256

Martínez-González MÁ, Hershey MS, Zazpe I, Trichopoulou A (2017) Transferability of the Mediterranean diet to non-Mediterranean countries. What is and what is not the Mediterranean diet. Nutrients 9(11):1226

Moreno LA, Meyer R, Donovan SM, Goulet O, Haines J, Kok FJ, Van't Veer P (2022) Perspective: striking a balance between planetary and human health—is there a path forward? Adv Nutr 13(2):355–375

Nassif F, Tanji A (2013) 3. Gathered food plants in Morocco the long-forgotten species in ethnobotanical research by Fatima Nassif and Abbes Tanji. Life Sci Leaflets 37:17

Novaczek I (2001) A guide to the common edible and medicinal sea plants of the Pacific Islands, vol 3. University of the South Pacific, Suva, p 40

Onwezen MC, Bouwman EP, Reinders MJ, Dagevos H (2021) A systematic review on consumer acceptance of alternative proteins: pulses, algae, insects, plant-based meat alternatives, and cultured meat. Appetite 159:105058

Ostfeld RJ (2017) Definition of a plant-based diet and overview of this special issue. J Geriatric Cardiol: JGC 14(5):315

Pawera L, Lipoeto NI, Khomsan A, Zuhud EAD (2018) Food plants of Minang and Mandailing communities in Pasaman District, West Sumatra. In: A community guidebook on biodiversity for nutrition and health, p 9

Pawera L, Khomsan A, Zuhud EAM, Hunter D, Ickowitz A, Polesny Z (2020) Wild food plants and trends in their use: from knowledge and perceptions to drivers of change in West Sumatra, Indonesia. Foods:1240

Penafiel D, Lachat C, Espinel R, Van Damme P, Kolsteren P (2011) A systematic review on the contributions of edible plant and animal biodiversity to human diets. EcoHealth 8(3):381–399

Penafiel D, Vanhove W, Espinel RL, Van Damme P (2019) Food biodiversity includes both locally cultivated and wild food species in Guasaganda, Central Ecuador. J Ethnic Foods 6(1):25

Pettersen BJ, Anousheh R, Fan J, Jaceldo-Siegl K, Fraser GE (2012) Vegetarian diets and blood pressure among white subjects: results from the Adventist Health Study-2 (AHS-2). Public Health Nutr 15(10):1909–1916

Popkin BM, Corvalan C, Grummer-Strawn LM (2020) Dynamics of the double burden of malnutrition and the changing nutrition reality. Lancet 395(10217):65–74

Pretty JN, Ball AS, Lang T, Morison JI (2005) Farm costs and food miles: an assessment of the full cost of the UK weekly food basket. Food Policy 30(1):1–19

Qian F, Liu G, Hu FB, Bhupathiraju SN, Sun Q (2019) Association between plant-based dietary patterns and risk of type 2 diabetes: a systematic review and meta-analysis. JAMA Intern Med 179(10):1335–1344

Quek J, Lim G, Lim WH, Ng CH, So WZ, Toh J, Pan XH, Chin YH, Muthiah MD, Chan SP, Foo RS (2021) The association of plant-based diet with cardiovascular disease and mortality: a meta-analysis and systematic review of prospect cohort studies. Front Cardiovasc Med 8:756810

Rivera D, Obon C, Heinrich M, Inocencio C, Verdea A, Fajardoa J (2006) Plants–Ethnobotanical Investigations and Historical Development. Local Mediterranean Food Plants Nutraceuticals 59:18

Romanelli C, Cooper D, Campbell-Lendrum D, Maiero M, Karesh WB, Hunter D, Golden CD (2015) Connecting global priorities: biodiversity and human health: a state of knowledge review. World Health Organistion/Secretariat of the UN Convention on Biological Diversity

Roth GA, Mensah GA, Johnson CO, Addolorato G, Ammirati E, Baddour LM, Barengo NC, Beaton AZ, Benjamin EJ, Benziger CP, Bonny A (2020) Global burden of cardiovascular diseases and risk factors, 1990–2019: update from the GBD 2019 study. J Am Coll Cardiol 76(25):2982–3021

Sanders TAB (1999) The nutritional adequacy of plant-based diets. Proc Nutr Soc 58(2):265–269

Sarfo J, Keding GB, Boedecker J, Pawelzik E, Termote C (2020) The impact of local agrobiodiversity and food interventions on cost, nutritional adequacy, and affordability of women and children's diet in northern Kenya: a modeling exercise. Front Nutr 7:129

Satter MMA, Khan MMRL, Jabin SA, Abedin N, Islam MF, Shaha B (2016) Nutritional quality and safety aspects of wild vegetables consume in Bangladesh. Asian Pac J Trop Biomed 6(2):125–131

Schwab K, Malleret T (2020) The great reset. In: World economic forum, Geneva, vol 22. Forum Publishing, Cologny/Geneva
Singh B, Sultan P, Hassan QP, Gairola S, Bedi YS (2016) Ethnobotany, traditional knowledge, and diversity of wild edible plants and fungi: a case study in the Bandipora district of Kashmir Himalaya, India. J Herbs Spices Med Plants 22(3):247–278
Springmann M, Clark M, Mason-D'Croz D, Wiebe K, Bodirsky BL, Lassaletta L, De Vries W, Vermeulen SJ, Herrero M, Carlson KM, Jonell M (2018) Options for keeping the food system within environmental limits. Nature 562(7728):519–525
Steffen W, Richardson K, Rockström J, Cornell SE, Fetzer I, Bennett EM, Biggs R, Carpenter SR, De Vries W, De Wit CA, Folke C (2015) Planetary boundaries: guiding human development on a changing planet. Science 347(6223):1259855
Stocker T (ed) (2014) Climate change 2013: the physical science basis: working group I contribution to the fifth assessment report of the Intergovernmental Panel on Climate Change. Cambridge university press
Sun H, Saeedi P, Karuranga S, Pinkepank M, Ogurtsova K, Duncan BB, Stein C, Basit A, Chan JC, Mbanya JC, Pavkov ME (2022) IDF diabetes atlas: global, regional and country-level diabetes prevalence estimates for 2021 and projections for 2045. Diabetes Res Clin Pract 183:109119
Tansey G (2012) Farming, food and global rules. In: The future control of food. Routledge, pp 3–24
Tilman D, Clark M (2014) Global diets link environmental sustainability and human health. Nature 515(7528):518–522
Tonstad S, Butler T, Yan R, Fraser GE (2009) Type of vegetarian diet, body weight, and prevalence of type 2 diabetes. Diabetes Care 32(5):791–796
Tonstad S, Stewart K, Oda K, Batech M, Herring RP, Fraser GE (2013) Vegetarian diets and incidence of diabetes in the adventist health study-2. Nutr Metab Cardiovasc Dis 23(4):292–299
Turner-McGrievy G, Mandes T, Crimarco A (2017) A plant-based diet for overweight and obesity prevention and treatment. J Geriatr Cardiol: JGC 14(5):369
Ulian T, Flores C, Lira R, Mamatsharaga A, Mogotsi KK, Muthoka P, Ngwako S, Nyamongo DO, Omondi W, Sanogo AK, Sanogo S (2019) Wild plants for a sustainable future. Kew Publishing, p 486
Ulian T, Sacandé M, Hudson A, Mattana E (2017) Conservation of indigenous plants to support community livelihoods: the MGU–useful plants project. J Environ Plan Manag 60(4):668–683
Vermeulen SJ, Campbell BM, Ingram JS (2012) Climate change and food systems. Annu Rev Environ Resour 37:195–222
WHO (2015) Connecting global priorities: biodiversity and human health: a state of knowledge review. WHO
Willett W, Rockström J, Loken B, Springmann M, Lang T, Vermeulen S, Garnett T, Tilman D, DeClerck F, Wood A, Jonell M (2019) Food in the Anthropocene: the EAT–Lancet Commission on healthy diets from sustainable food systems. Lancet 393(10170):447–492
Williams KA, Patel H (2017) Healthy plant-based diet: what does it really mean? J Am Coll Cardiol 70(4):423–425
Woodside J, Young IS, McKinley MC (2022) Culturally adapting the Mediterranean diet pattern–a way of promoting more 'sustainable'dietary change? Br J Nutr 128(4):693–703
Yokoyama Y, Barnard ND, Levin SM, Watanabe M (2014) Vegetarian diets and glycemic control in diabetes: a systematic review and meta-analysis. Cardiovasc Diagn Ther 4(5):373
Zhou B, Lu Y, Hajifathalian K, Bentham J, Di Cesare M, Danaei G, Bixby H, Cowan MJ, Ali MK, Taddei C, Lo WC (2016) Worldwide trends in diabetes since 1980: a pooled analysis of 751 population-based studies with 4· 4 million participants. Lancet 387(10027):1513–1530

# Chapter 4
# Innovations in Plant-Based Food Production

**Abstract** A change from conventional methods to more conscientious alternatives is needed in order to provide healthy, secure and easily accessible food with the least possible negative effects on the planet. The significance of sustainable food production technology in supplying the world's food needs while tackling worldwide environmental issues will be addressed in this chapter article. Precision farming, vertical farming, fogponics, hydroponics, and aquaponics are among some of the important technologies. By collecting data on soil and water fluctuations and improving techniques like planting, fertilizer and irrigation, precision farming leverages technology to increase farming efficiency. Additionally, nanotechnology is becoming recognized as another technique that has the potential to restructure agriculture by offering effective and sustainable solutions for resistance to diseases, crop development and nutrient regulation. Lastly, plant-based, cultured meat, dairy products, mycoprotein and microalgae are examples of future food trends that have surfaced as a sustainable substitute for conventional meat production. It is essential to use the aforementioned technology into farming methods in order to accomplish several Sustainable Development Goals.

**Keywords** Aeroponics · Aquaponics · Fogponics · Mycoproteins · Nanotechnology · Plant based meat

## 4.1 Introduction

One of the main pillars of our civilization, agriculture has always changed to accommodate new technologies and societal developments. Food production has been vital for human growth and civilization since the beginning of crop cultivation about 10,000 years ago and continues to this day in the form of extremely sophisticated agricultural systems (Lal et al. 2007). However, a number of environmental problems, such as water pollution, soil erosion, ecological degradation and greenhouse gas emissions, have been exacerbated by both conventional and industrial agricultural practices. Food production increased significantly during the Green Revolution



N. Patel et al., *The Future of Food: Exploring Edible Plants for a Sustainable Diet*,
https://doi.org/10.1007/978-3-032-15027-1_4

in the middle of the twentieth century, mostly as a result of the introduction of highly productive crop types, artificial fertilizers, and pesticides (Martínez-Romero et al. 2020). It was a comprehensive program designed to increase food production in underdeveloped nations by utilizing more sophisticated farming practices. The strategy raised questions about its long-term sustainability because, increasing food production by more than 1000% in certain places, it also had detrimental effects on the environment, higher manufacturing costs and decreased biodiversity (Çakmakçı et al. 2023). Even if it reduces hunger in several impoverished nations, more sustainable approaches must be used in future initiatives. The effective cultivation of safe, high-quality agricultural products in a manner that preserves and enhances the natural surroundings, socioeconomic status of farmers, their workers and local communities, as well as the health and welfare of all farmed species, is known as sustainable agriculture (Marvin et al. 2022).

Producing food sustainably is becoming essential rather than optional. This method replaces existing (conventional), frequently destructive methods of agriculture with more conscientious, alternative methods that lessen their detrimental effects on the environment. As a result, agriculture is essential for improving food security, quality, and affordability. Nevertheless, considering the projected increase in future global food consumption, it is unclear if global agriculture will eventually be able to fulfill this obstacle (Rockstorm et al. 2020). Moreover, ecological changes profoundly affect food supply which additionally loads the concern for our prospective future. The climatic resilience of agricultural production is the capacity to minimize losses brought on by climate change and regain or even increase food yields. In general, increasing crop production and the amount of land utilized for farming are two possible ways to reduce hunger (Pawlak and Kołodziejczak 2020). It is equally important to encourage investments in extension services and agricultural infrastructure, particularly in rural areas. Current knowledge and technology may not be sufficient in less developed and developing countries to produce the required quantity of food both today and in the future. The 65% to 75% of the underprivileged who reside in rural areas get their food from smallholder agriculture, which comprises small-scale cultivators, ranchers, forest maintainers, and fishermen who oversee areas ranging from less than one hectare to ten hectares (Michler et al. 2019). This demonstrates the necessity of increasing agricultural research funding, especially in emerging nations, in order to raise output per agricultural worker and per unit of land. Additionally, the agricultural industry is greatly impacted by price variations, political unrest, and natural disasters, which are mostly brought on by climate change. Furthermore, as carbon emissions rise and affect the dependability of our water sources, climate change will make it more crucial than ever for irrigated agriculture to discover sustainable and profitable solutions (Walls et al. 2019).

In many regions of the world, rising temperatures has already had a major impact on and decreased food production. A more thorough plan for enhancing global nutrition would be provided by a holistic approach that takes these variables into account in addition to food safety, availability, and accessibility (Lipper et al. 2014). Therefore, in order to boost productivity while improving resistance to weather shocks and lowering negative externalities, prominent organizations and governments promote the use of climate-adaptive agricultural technologies. Hence, the present chapter thoroughly examines a number of cutting-edge technologies that have the potential to develop more environmentally friendly and ecologically sound agricultural systems, ranging from nanotechnology applications to precision agriculture, hydroponics, aquaponics and vertical farming.

## 4.2 Future Forward Technologies in Sustainable and Plant Based Food Production

In order to address the threats that climate change poses to our food systems, future forward technological advancements are essential. From cultivating crops to harvesting and post-harvest processing these technologies could improve agricultural production. Resilience, sustainability and efficiency at every stage.

### *4.2.1 Vertical Farming*

In order to meet the growing global population and the enormous demand for food, the farming sector is under pressure to boost production. By 2050, there may be nearly nine billion individuals on the planet, most of whom would reside in cities, according to projections (Beacham et al. 2019). Due to the constraints associated with traditional farming in the areas of both land accessibility and profitability, urbanization and infrastructural development may result in the loss of productive agricultural land, mandating the growth of innovative food production techniques. There are two types of vertical farming systems: systems that produce crops on an inclined surface and multi-level conventional horizontal growth platforms (Xi et al. 2022).

The primary advantage of NASA's research is the establishment of vertical farms. The nutrient film approach, for instance, has been embraced by potato farmers and has produced unprecedented yields. They also develop and execute innovative techniques. Despite being mainly utilized for leaf crops; some fruit crops can benefit from nutrient film cultivation. Recirculating water and nutrients via growing passages makes it achievable to build small devices to feed raising vegetative plants at home. Because of NASA's groundbreaking research, the existing $2 billion international vertical gardening/farming market is expected to grow to $7 billion by 2025 (Tilman et al. 2011). Industries have made great use of NASA's research on the best growing conditions for different plants to improve their cultivation methods. NASA found 15 plants that can be cultivated hydroponically and are essential for sustaining life, offering a practical way to feed humans on long-haul space journeys. Significantly, additional organizations establish urban vertical farms that offer local fresh food using NASA's information-driven models and computational intelligence to control production cycles (Suresh et al. 2017; Patnaik et al. 2020).

Due to soil-free cultivation methods, vertical systems minimize the demand for traditional agricultural land that is appropriate for urban areas. Larger systems are thought to be more sustainable, which influences customer acceptance of these systems. These systems do have certain drawbacks, though. Multiple growth surface levels make food production systems technically difficult and costly; these systems usually involve greenhouses with controlled environments that use hydroponics or aquaponics (Van Gerrewey et al. 2021). As a result, a thorough technical approach must be taken, considering factors such as site selection, construction, energy conservation, crop sustenance, illumination and growth systems. Even though vertical farming has demonstrated potential for certain important crops, additional research is required to maximize its technical and financial features (Jürkenbeck et al. 2019). Because of the paucity of existing research on vertical farming, further studies are required to increase production, minimize system costs, and engage prospective consumers. Currently, a huge number of independent start-ups are at the forefront of vertical farming, which is mostly fueled by industry. These systems are also not standardized because every organization uses a different strategy. The majority of the information on crop yields and the viability of vertical farming that is now available is not publicly available and is frequently based on conjecture or promotional materials rather than actual research (Benke and Tomkins 2017).

Vertical gardening has additional constraints. Even if vertical indoor farms offer a workable answer for food security, more crops ought to be cultivated indoors. These days, green leafy vegetables, culinary herbs, and berries like strawberries are among the most widely propagated plants. This range can be increased, though, by utilizing other technologies (Pinstrup-Andersen 2018). The emergence and blooming of tomatoes and ground cherries, for instance, have been altered for indoor culture through the use gene editing technologies. Targeting on the regulator of stem length of tomato in addition to developing a trait-stacking technique involving mutations for compressed shoots, quick flowering and premature growth termination, Solanaceae crop varieties were altered (Kwon et al. 2020). Plant characteristics were modified to accommodate indoor cultivation by gene manipulation and the use of CRISPR-Cas9 methods. According to the investigators, closely related species may need distinct genetic strategies in order to develop successfully on urban farms (Fabi et al. 2021). When combined with the right technologies, indoor vertical farming techniques can result in year-round harvesting, reduced water usage, and increased yields while avoiding the need for heavy pesticide use. However, the expenses of establishing and maintaining farm operations inside continue to be an obstacle, particularly when it comes to energy. In order to support growing urban populations in the face of ecological issues and make urban agriculture more cost-effective, future breakthroughs may require multifaceted study that takes into account genomics, agronomics, and engineering (Vishnoi and Goel 2024).

### 4.2.2 *Aquaponics*

Soilless farming uses effective water-saving strategies to address major issues in traditional farming, such as pollution, diseases of plants, and diet. Aquaponics is one among multiple techniques used to enable soilless agriculture, which is currently proven to be a profitable and environmentally friendly option. Even if the initial manufacturing costs are quite substantial, they could become more efficient and less expensive with more research. Aquaponics is a system that combines both aquaculture and hydroponics to produce fish and crops in an uncommon way. By using the waste products as a source of nutrients, the plants remove hazardous materials from the fish's microenvironment (Rakocy 2012). Aquaponics is being developed as a truly ecologically sound agricultural strategy by researchers and suppliers since the waste products of one biological system (fish) may be used as nutrients for another (plants). The byproducts of combining fish and plants produce a number of commodities. Water that has undergone both mechanical and biological filtration and recirculation can be reused to produce nutritious regional food that may stimulate the regional economy (Lennard and Goddek 2019).

Once solid waste has been removed by a swirl hydraulic filter, the roots of the plants collect nutrients from fish excrement in the resulting water and use them as fertilizer to nourish the plants hydroponically. In exchange, hydroponics removes phosphate, nitrates, nitrites, and ammonia by acting as a biological filter. In a closed cycle, the water can then be recycled back into the tanks that keep the fish clean and fresh. In the gravel, nitrifying bacteria proliferate as plant roots carry out nutrient cycling. The crucial component of the nitrification process that converts fish waste into plant-suitable nutrients are the nitrifying bacteria. In aquaponics systems, the process of nitrification removes nitrite and ammonia, which are harmful to fish and plants alike, while also supplying nutrients for the plants. Both nitrite oxidizing bacteria and ammonia oxidizing bacteria work to break down the ammonia in fish waste into nitrite, which is subsequently changed into nitrate by nitrite oxidizing bacteria. Nitrate is the primary fertilizer used by plants for growth purposes (Bernstein 2023). The most significant function of the aquaponics system is its closed water reuse-cycled system, which allows farms with water constraints to produce fresh veggies and fish protein. Despite its widespread recognition, aquaponics is still not widely used in developing nations. However, it is becoming more popular in nations like South Africa and Egypt, where high adoption rates are improving food security. Notably, aquaponics systems constructed locally in West Africa have proven to be more economically viable than those that use imported materials. Even if they are more expensive initially, locally produced aquaponics systems have an enormous amount of prospects for the future (Baganz et al. 2022).

Aquaponics meets worldwide food needs by using land and water/nutrients more efficiently than conventional farming. However, they encounter difficulties such comparatively high energy use and disputes regarding their organic identity. Aquaponics, yields almost twice the economic value over a month compared to hydroponics, which solely produces vegetables, according to a comparative life cycle assessment conducted in the United States. Because its products are more valuable overall, aquaponics had a 45% lower endpoint ecological footprint than hydroponics (Chen et al. 2020; Taha et al. 2022). The manufacturing of fertilizer and fish feed has an extra influence on the environment, although the primary causes are the power used in greenhouse operations and the water management in both systems. Making the switch to alternative energy sources like solar or wind could make things more ecologically friendly.

### *4.2.3 Hydroponics*

Approximately 3.5% of the surface area used for growing vegetables worldwide is being utilized in greenhouses and tunnels employing soilless agricultural methods based on hydroponic solutions. This substantial dispersion at the field scale clearly demonstrates that this production method has other benefits beyond its more effective use of water and other nutritional resources (Kinoshita et al. 2016). In fact, there are numerous instances where hydroponic solutions can be effectively utilized for bio fortification initiatives involving trace elements (such as calcium (Ca), silicon (Si), iodine (I), and selenium (Se), to enhance vegetable quality and shelf life in accordance with customer preferences and market trends. An entirely novel model in the commercial arena, known as Industry 4.0 or Smart production, is developing. It relies on Internet of Things and cyber-physical production systems. It is noteworthy that manufacturing of vegetables hydroponically in small, well regulated spaces would lessen the most significant distinctions between industrial and agricultural processes, which will enhance quality control. However, a series of extra monitoring instruments designed especially for the hydroponic-based growth method are required to accomplish this purpose (Rodriguez-Ortega et al. 2017).

Hydroponically produced plants and soil-grown plants are physiologically identical. Before being accessible to plants, the natural and synthetic components of soil must break down into inorganic elements. After adhering to the soil particles, these parts are transferred into the soil solution, where plants assimilate them. The plant roots in hydroponics are hydrated using a nutrient solution that includes the necessary components. The succeeding mechanisms by which the plant absorbs minerals are identical. The drip system, nutrient film technology, and deep-water culture are all parts of hydroponic systems. A deep-water culture system involves placing plant roots in net containers with a neutral substrate and suspending them in nutrient-rich water. A breathing stone supplies air straight to the roots (Sardare and Admane 2013). In contrast, a drip system uses drip nozzles to slowly administer the nutritional solution to the plants. In contrast, the nutrition film approach circulates the nutrient solution via raised tunnels where the plants are positioned using pumps, enabling better absorption of nutrients and increased root aeration (Jan et al. 2020).

By allowing for exact control over the nutrient supply, hydroponics offers a major nutritional benefit to plants. Hydroponics promotes optimal nutrient absorption by combining a balanced fertilizer solution with an effective delivery system. Compared to conventional cultivation, this direct access to essential nutrients speeds up development, resulting in healthier plants with larger harvests. Furthermore, the product's taste, size, and nutritional value may all be optimized with exact control over the components. Because hydroponics uses closed-loop systems, it also offers a significant benefit over traditional soil-based agriculture, which has a high rate of water evaporation. By capturing and recycling the nutrient solution, these devices reduce water waste by providing nutrients to plant roots. Hydroponic systems may still require significant amounts of electricity, even with advancements in energy-efficient technologies. The operation of temperature control systems, artificial lighting, and other mechanical components raises overall energy usage. High installation costs and a shortage of workers with the technical know-how required to generate the best possible crop production and quality are further issues. This technique is used to cultivate a variety of crops in huge glasshouses, including leafy greens like lettuce, herbs, tomatoes, and peppers. It may consist of substrate blocks made of rock wool or comparable materials that form a matrix for plant roots and are drip-fed a carefully balanced solution of nutrients and water. Another option is to use a thin coating of nutrient solution in the area around the roots or to grow plants in rafts that float on the top of sheets of nutrient solution. These systems frequently include fertilizer solution recirculation to guarantee the ideal nutritional composition required for plant growth (Khatri et al. 2024).

### *4.2.4 Aeroponics*

Aeroponics is a progressive form of hydroponics that involves growing plants without soil or other aggregation media in a mist environment. This novel growing method maximizes oxygen exposure and nutrient digestion by allowing plant roots to directly absorb nutrients from a thin mist. Increased air exposure speeds up root development and gives the plant more control over its environment. When compared to traditional agricultural methods, aeroponic systems may produce higher yields, faster growth rates, and improved water efficiency (Singh 2023). Additionally, because to its soil-independent nature, aeroponics reduces the risk of soil-borne pests and diseases, making it the best option for urban agriculture with limited area and perhaps for farming on other planets. Aeroponic chambers, misting nozzles, nutrient reservoirs, submersible pumps, cycle timers, root supports, pressure gauges, drainage systems, air filtration systems, lighting systems, and environmental controls are some of the parts of aeroponics (Ragaveena et al. 2021).

With an advantage in water conservation, aeroponics offers all the advantages of hydroponics. Aeroponics uses 70–90% less than hydroponics. Furthermore, because the roots are in contact with the nutritional solution mist directly, they receive more oxygen and are able to grow without limitations. Additionally, this stops pathogen growth on the roots and could lower pesticide control costs. Furthermore, high root zone temperatures in other hydroponic systems limit root growth and result crop nutrient shortages, which reduce photosynthetic efficiency. In aeroponic systems, root zone cooling reverses this inhibition (Fussy and Papenbrock 2022). However, the disadvantage is that crop failure would result from any system failure that halted fertilizer supply. Therefore, routine inspections are required. Furthermore, in order to pump the nutritious solution through misting devices, this system primarily depends on a power supply.

### 4.2.5 Fogponics

Fogponics is a sophisticated type of aeroponics that encloses hovering plant roots by transferring nutrients and oxygen via vaporizing water. Fogponics uses a foliar feeding technique and a 5–30 μm mist inside the root chamber, both of which follow the main principles of aeroponics. Using an ultrasonic atomizer, the nutrition solution is vaporized onto the roots. An electrical instrument called an ultrasonic atomizer turns water into a mist of minuscule droplets, measuring around 4 microns (Uddin and Suliaman 2021). Particle size indicates faster absorption for plants. Because these droplets are so minute, the root's pores may absorb them instantly. This method works very well in the horticultural sector. An additional advantage of employing a fogponics system over a conventional hydroponics system is that the crops may still support a huge plant while requiring less energy for root growth and weight (Sadar and Chyon 2020).

Because fogponics eliminates the need for agricultural chemical products and conserves water, it is a boon to the farming system. This approach favors organic farming methods, which are more profitable than the inorganic food process. Since organic food is known for its flavor and freshness, a significant amount of safe, pesticide-free food can be produced. Numerous crops can be cultivated in a controlled environment that gives them exactly what they need to produce food in the most cost-effective manner (Venkatesh et al. 2020).

### 4.2.6 Precision Farming

Precision farming increases farming productivity through the use of technology. In order to optimize farming techniques like planting, fertilizing, and irrigation, it entails collecting data on differences that occur within a field (such as the state of the soil, crop health). However, the quality of the information may suffer if data from several sources are combined. Concerns regarding safety and protection are also raised by the availability of the collected data (Cisternas et al. 2020). Furthermore, precision agriculture can lower environmental impacts, increase yields, and use fewer inputs by utilizing remote sensing, the Global Positioning System, and other technologies. Crop health monitoring solutions based on the Internet of Things may distinguish between both healthy and diseased crops with the help of wireless sensor networks and satellite imaging platforms. Crop productivity can be increased, irregularities may be mitigated, and input costs can be reduced with precision farming. Precision farmers can identify both spatial and temporal differences in their resources by using information technologies to gather a variety of data about the land. This makes it easier to precisely apply treatments where they are required (Srivastav et al. 2021). Expenditure can be substantially decreased, for instance, by applying fertilizers based on soil sample and yield data analysis. Additionally, it increases the effectiveness of harvesting techniques and water resource management (Bayabil et al. 2022).

By addressing the technical, financial, and ecological facets of farming operations, the integration of information technology also facilitates the execution of smart choices. Precision farming uses a variety of tools to gather location-specific data and implement appropriate management methods, as opposed to managing an entire field based on a notional average condition that might not even exist anywhere in the field (Finger et al. 2019). Precision farming is still not widely adopted despite its advantages. Germany and the United States have conducted the majority of the research on its implementation. Though they have been around since the 1990s, precision farming methods are more common in industrialized nations, where adoption is still relatively low (Raj et al. 2022).

## 4.3 Genetically Modified Crops: A Biotechnological Arsenal Against Global Food Challenges

Crops with desired characteristics, such resistance to pests and diseases, drought, or higher nutritional value, can be generated via genetic modification. Genetically modified/engineered crops have the potential to be a potent instrument in enhancing the adaptability of our food systems to the impacts of climate change. These crops have demonstrated a great deal of promise in reducing greenhouse gas emissions from agriculture, both production emissions and carbon opportunity cost (Atlin et al. 2017). These crops can reduce the need for land expansion by increasing crop yields, protecting native vegetation and lowering emissions associated with land usage, including fertilizer and energy input emissions. Additionally, crops' capacity to endure harsh environmental conditions can be improved by breeding them to withstand drought and high temperatures (Kumar et al. 2020).

Numerous scientific studies have shown the benefits of genetically engineered crops, despite the continuous public discussion about their possible hazards. These advantages include higher agricultural yields, more profitable farms, and occasionally less use of pesticides. By enabling low-tillage farming methods, various genetically engineered crops have also shown promise in lowering greenhouse gas emissions and encouraging soil carbon absorption. The crop productivity enhancements provided by these crops could help avoid extra land usage and hence lower CO2 emissions as the demand for agricultural products rises globally (Noack et al. 2024). The European Parliament recently endorsed its stance in negotiations to promote farmers' green transition by utilizing new genetic approaches in agriculture. By applying genetic approaches to make crops durable, pest-resistant, and higher productive while using fewer fertilizers and pesticides, agricultural systems should become more sustainable. Even though complicated alterations would be subject to stronger regulations, using modified crops in organic farming will still be forbidden. Furthermore, it is recommended that all patents pertaining to this topic be prohibited in order to safeguard farmers (Sandhu et al. 2025).

## 4.4 Nanotechnological Revolution

Nanotechnology is contributing significantly to the advancement of agriculture by providing novel solutions for productive and ecologically sound farming. By facilitating the accurate distribution of agrochemicals, it enhances plant growth, disease resistance and nutrient management. Products with Nano encapsulation make it possible to apply pesticides, insecticides, and herbicides more precisely and efficiently, which promotes environmentally friendly farming methods (Pulizzi 2019). Furthermore, postharvest applications of nanotechnology reliably preserve the quality, freshness, and shelf life of crops that are stored while averting infections.

Synthetic nanochemicals, nanofungicides, and nanofertilizers are becoming more and more popular as a sustainable way to treat and prevent plant diseases. This technique lowers environmental risks and agricultural inputs while increasing efficacy. Through boosting intrinsic defense mechanisms, nanoparticles like copper and silver shield crops like wheat from infections. Iron oxide nanoparticles have shown promise for crop protection in stressful situations by being utilized to lessen the harmful effects of cadmium in wheat (Pramanik et al. 2020). Additionally, rice is protected from a variety of diseases by green-synthesised ZnO nanoparticles. When applied to pests such as *Sitophilus oryzae*, silica nanoparticles have demonstrated entomotoxic effects (Jurić et al. 2024).

Nanoparticles has been utilized in fertilizer development to supply nutrients to plants in an effective and targeted manner. For example, silver nanoparticles controlled by DNA on graphene oxide function as antibacterial agents against the plant pathogen *Xanthomonas perforans* (Ogunyemi et al. 2019). Additionally, mesoporous silica nanoparticles improve photosynthesis and seedling growth in wheat and lupin, whereas carbon nanotubes improve tomato seed germination and plant growth. These instances highlight how important nanotechnology is to promoting more effective and sustainable farming methods.

# 4.5 Future Food Trends

## 4.5.1 *Lab Engineered Meat*

A greater number of individuals are interested in lab engineered meat also known as cultured meat, which is prepared by growing modified muscle tissue in a lab. It has the potential to have a less negative environmental impact than conventional animal husbandry, but it is still in its infancy and will not yet reach its full potential in terms of impact and optimization. Although cultured meat has a high nutritional potential and is biologically identical to regular meat, customer acceptance of it varies (Kumar et al. 2021). The acceptance of cultured meat by consumers is lower than that of plant-based meat substitutes, pulses, or algae. At the moment, regulatory obstacles and high production costs prevent scaling.

It's important to note that the culture medium accounts for a sizable amount of the expenditures associated with producing cultured meat, from 55% to 95% of marginal costs. This media is expensive since it needs components that are specified by chemicals and do not involve animals. Using food-grade materials, optimizing growth factors, or modifying cell lines are some ways to lower these expenses (Tuomisto and Teixeira de Mattos 2011). By localizing growth factor release, simulating non-covalent sequestration or directly integrating mimetics inside the scaffold, engineered biomaterials may improve growth factor efficiency and reduce costs.

There is still a dearth of information on the qualitative characteristics of cultured meat because the area is new and research is concentrated in industry companies rather than academia. Connecting the dots between biomedical engineering and meat biology may provide novel solutions to production problems. Market acceptance and integration into current meat manufacturing pipelines could be accelerated by taking these limits into account early in the design phase (Roy et al. 2021).

## 4.5.2 *Microalgae and Mycoprotein*

Aquatic unicellular organisms known as microalgae, which are essential to world ecosystems, are showing promise as an alternate source of protein. They are vital for photosynthesis, which transforms sunlight into organic matter. They also play a major part in the generation of oxygen and the fixation of $CO_2$ (Dolganyuk et al. 2020). Compared to traditional agriculture, microalgae farming has a smaller environmental impact, particularly in terms of land use and greenhouse gas emissions. They flourish in a variety of habitats, such as industrial wastewater and saltwater, without posing a threat to biodiversity or vying for arable land. Because of their versatility, microalgae could be a future food source to help with the world's food deficit.

Microalgae are higher in proteins than most conventional plant- or animal-based diets, frequently making up more than 50% of the dry content. Due to intricate technical problems in growth, harvesting, and downstream processes, their protein digestibility ranges from 57% to 90%, which is somewhat lower than that of insect proteins (Chen et al. 2022). This poses difficulties for purifying and production scaling. The potential for using microalgae cultivated in anaerobically digested wastewater to produce protein-dense products is enormous. The potential to produce 2 kg of microalgae per cubic meter of wastewater at a cost varying from 1 to 8 USD per kg, depending on the scale, was identified by a techno-economic study conducted in Western Australia (Lucakova et al. 2022).

This procedure may yield protein isolates, concentrated product, and animal feed at prices of 30 USD, 25 USD, and 5 USD per kilogram, respectively, at an annual production scale of 1600 cubic meters. Scaling up manufacturing could produce achievable results even though it is not now economically viable. However, additional wastewater collection costs must be taken into account (Smetana et al. 2017). According to the findings, microalgae produced at quantities ranging from 180 to 1600 cubic meters per day in Western Australia could be produced for a minimum of 1.70–8.05 USD per kg. Although the expenses of moving wastewater to a centralized facility needs to be further assessed where a centralized plant would be a good solution (Yang et al. 2024).

However, based on life cycle studies, studies on sustainability shows that both autotrophic and heterotrophic microalgae species have a greater environmental impact than conventional protein sources (Mellor et al. 2022). Technological developments aimed at improving the productivity, quality, and protein content of microalgae farming hold promise for streamlining the harvesting and processing phases, which could counteract negative effects on the environment and the economy. The techno-functional qualities of foods containing microalgae proteins are affected by strong flavors and colors, and consumers' awareness of algae as a food source is still low. Establishing algal protein as an ecologically friendly and practical substitute for conventional food sources requires overcoming these challenges (Rizwan et al. 2018).

On the other hand, Mycoprotein's unique fibrous structure and functional dietary qualities have led to its growing recognition as a sustainable meat substitute. Since, middle of the twentieth century, there has been interest in employing fungal cells and materials as protein sources rather than proteins generated from cattle (Khan et al. 2024). However, there are still issues with mycoprotein products' mass production, texture, sustainability, and customer acceptability. After 15 years of research and development, QuornTM, the most well-known mycoprotein product on the market, emerged on sale in 1985. The filamentous fungus *Fusarium venenatum* is used in a pressure cycle fermenter to generate a variety of goods, such as alternative chicken patties, sausages, and burgers. What emerges is mycoprotein products that contain 60–90% protein on a dry basis. The fungal mycelium's branching structure gives it an appearance comparable to meat by imitating muscle fibers (Hashempour-Baltork et al. 2020).

With the exception of meat and poultry products, mycoprotein has been deemed generally recognized as safe by the US Food and Drug Administration since 2001. It is relatively low in fat, high in protein and rich in fiber, with β-glucan and chitin. Mycoprotein has a net protein utilization value that is similar to that of milk and contains all of the essential amino acids. It also supplies vital minerals including calcium, phosphorus, zinc, iron, salt, selenium, manganese, and Vitamin B2 (Barzee et al. 2021). Investigation demonstrates that supplementing mycoprotein to the diet can improve lipid profiles while lowering energy intake in lean, obese and overweight people. Despite mycoproteins'higher fiber and lower calorie content, it is not advised for kids (under 3 years) because of their high energy needs during rapid growth. The manufacturing of mycoproteins uses minimal water and produces less greenhouse gas emissions, than the production of commercial animal proteins. These characteristics demonstrates the potential of mycoproteins as a sustainable substitute for other forms of protein in food systems of the future (Ahmad et al. 2022).

### 4.5.3 Plant Derived Meat and Dairy Products

Plant derived meat (PDM) is made to resemble meat for "meat lovers" or flexitarians as well as vegetarians. Analyzing and choosing suitable plants and other raw materials are necessary steps in the development of PDM. For instance, modified legume protein isolates that have undergone hydrostatic pressure processing and are incorporated into PDMs enhance functional characteristics throughout the product formulation process (Bellone et al. 2022). A mixture of legumes, such as pea and oat protein, would enhance the amino acid profile, wettability, and stability across a broad temperature range in addition to supplying the necessary protein amount for desirable texture in projected meat substitutes. However, as the proportion of oat protein in the formulation rises, consequently increases the stiffness and moisture content of substitute meat products. To enhance the sensory qualities of PDMs, additional nutrients such fats, starches, and lecithin are added. When compared to burger formulations made without lecithin, those that have it are juicier and tastier, uniformly brown, and sizzle when baked (Kyriakopoulou et al. 2019). Additionally, Herbacel, a commercial citrus fiber, has been shown to enhance final product texture and freeze-thaw stability while reducing cooking and frying loss (Tachie et al. 2023).

Dairy substitutes made from plants currently account for 7% of the milk market, and this percentage is projected to increase by 8% by 2031. Plant-based dairy products have unique flavors and are made from milk that comes from legumes (soy, chickpea, bambara), seeds (sunflower, hemp), nuts (hazel, coconut), pseudo cereals (quinoa), and cereals (oat). Diversifying plant-based milk has been suggested as a way to avoid taking advantage of the market for a certain product and to maximize nutritional benefits (Laila et al. 2021). Lactose-intolerant consumers are interested in plant-based dairy products since they are lactose-free. The primary objective of the plant-based dairy sector is to provide milk that has comparable nutritional and organoleptic qualities with standard milk. In order to make non-dairy milk, dairy products extract from the preferred plant source is mixed with additional ingredients like oil, emulsifying agents, thickeners (locust bean gum), additives (enzyme deactivators), flavors, preservatives and nutrients (tricalcium phosphate) (Tangyu et al. 2019). The mixture is then filtered, sterilized, and thoroughly homogenized to avoid separation over time. Yogurt, cheese, and milk are a few examples of commercially available plant-based dairy products that are already accessible to consumers. Customers can also at present purchase an industrial plant-based powdered milk that has been developed to make the finished product lighter for convenience and that they simply need to dissolve in water as needed (Aydar et al. 2020; Mozaffari and Nouman 2021).

### 4.5.4 Meadow Orchards and High Density Planting

Globally, intensive agriculture is acknowledged as a primary contributor to the decline of biodiversity. Changes are being demanded by national and international organizations and institutions to improve the total agricultural impact in a way that is more conservation-friendly and sustainable. Novel ways are needed to support the long-term provision of linked ecosystem processes and services, preserve life in and around agricultural areas, and avert an ecological collapse (Sánchez-Bayo and Wyckhuys 2019). Meadow orchards, another term for classic fruit orchards, are one type of these production methods. Up until the twentieth century, this traditional type of agroforestry was common throughout central Europe and was typified by sporadic, primarily high-stem, old-grown fruit trees with a grassland understory (Edo et al. 2024). The cultural value of meadow orchards and their capacity to offer a variety of ecosystem services and functions, on the other hand, have recently led to a rise in their acknowledgment. Furthermore, meadow orchards are especially rich in biodiversity due to the low intensity of grassland and availability of old fruit trees. The majority of research has compared meadow orchards with other land-use categories, like grasslands and forested regions, or with varying management intensities. Meadow orchards are consistently highlighted in these studies as having a remarkably high level of biodiversity. Only a small number of studies, however, compared intensive and meadow orchards, concentrating on individual species like birds and soil macro fauna (Forejt and Syrbe 2019). On the other hand, High Density Planting is the dense planting of smaller trees, followed by mechanical harvesting, chemical growth control, rootstock dwarfing, and pruning. The fundamental idea behind high-density planning is to maximize return per unit of input and national resources while utilizing vertical and horizontal space as efficiently as feasible. It also provides both short-term and long-term fruit crops with excellent productivity per unit area (Joshi 2020).

## 4.6 Hidden Barriers: What Holds Back Plant Based Diet Adoption

The effects of Plant derived foods vary widely, despite the fact that their use is being promoted and that their patterns have been linked to improved consumer and environmental health. For example, increased consumption of unhealthy plant derived diet items including refined grains, fruit juice, sugary drinks, potatoes and sweets/desserts was linked to higher scores on the unhealthy plant diet index (Musicus et al. 2022). Because of their propensity to form complexes with proteins and minerals, certain plant derived ingredients, like legumes and cereals, contain variable amounts of anti-nutrients like phytates, saponins, tannins, protease and amylase inhibitors. These compounds restrict the amount of the ingredient which can be utilized in formulation decreasing the digestion of proteins and the overall quality of nutrition, preventing the uptake of minerals, causing discomfort in the stomach, and becoming noxious when accumulated (Samtiya et al. 2020). Pulses are typically soaked for a few hours in order to reduce their anti-nutrient content and cooking time; however, certain consumers may find this inconvenient. Due to the inadequate levels of vitamins A, B2, B12 and D and the poor bioavailability of certain minerals (including calcium, zinc, iron, and iodine) plant derived foods must be critically examined and additional/alternative sources of these nutrients must be included (Szczebyło et al. 2020).

Many traditional vegetarian dishes, including tofu, can be prepared with less salt and oil. However, certain vegetable-based sausages and burgers on the market have higher levels of saturated fats and salt, which increases the amount of calories and salt that are consumed. Concerns have also been raised by certain customers regarding the high cost, limited availability in supermarkets, and potential harm to oral health caused by some plant-based dairy products (Tso and Forde 2021).

## 4.7 Conclusion

A comprehensive plan for enhancing global food security would need a holistic approach that takes into account food safety, availability, and accessibility. Current knowledge and technology, however, might not be sufficient in developing countries, underscoring the need for more funding for agricultural research. There are still many restrictions and issues with integrating the aforementioned technologies, even if they have the potential to lessen issues with environmental resilience and sustainable food production. Numerous issues are linked to complexity, adaptability, effectiveness, investments, legislation, the safety of food, and, lastly, consumer perception. While some of these issues can be resolved, many call for difficult fixes.

## References

Ahmad MI, Farooq S, Alhamoud Y, Li C, Zhang H (2022) A review on mycoprotein: history, nutritional composition, production methods, and health benefits. Trends Food Sci Technol 121:14–29

Atlin GN, Cairns JE, Das B (2017) Rapid breeding and varietal replacement are critical to adaptation of cropping systems in the developing world to climate change. Glob Food Sec 12:31–37

Aydar EF, Tutuncu S, Ozcelik B (2020) Plant-based milk substitutes: bioactive compounds, conventional and novel processes, bioavailability studies, and health effects. J Funct Foods 70:103975

Baganz GF, Junge R, Portella MC, Goddek S, Keesman KJ, Baganz D, Staaks G, Shaw C, Lohrberg F, Kloas W (2022) The aquaponic principle—It is all about coupling. Rev Aquac 14(1):252–264

Barzee TJ, Cao L, Pan Z, Zhang R (2021) Fungi for future foods. J Future Foods 1(1):25–37

Bayabil HK, Teshome FT, Li YC (2022) Emerging contaminants in soil and water. Front Environ Sci 10:873499

Beacham AM, Vickers LH, Monaghan JM (2019) Vertical farming: a summary of approaches to growing skywards. J Hortic Sci Biotechnol 94(3):277–283

Bellone F, Cinquegrani M, Nicotera R, Carullo N, Casarella A, Presta P, Andreucci M, Squadrito G, Mandraffino G, Prunestì M, Vocca C (2022) Role of vitamin K in chronic kidney disease: a focus on bone and cardiovascular health. Int J Mol Sci 23(9):5282

Benke K, Tomkins B (2017) Future food-production systems: vertical farming and controlled-environment agriculture. Sustain Sci Pract Policy 13(1):13–26

Bernstein S (2023) Aquaponic gardening: a step-by-step guide to raising vegetables and fish together. New Society Publishers

Çakmakçı R, Salık MA, Çakmakçı S (2023) Assessment and principles of environmentally sustainable food and agriculture systems. Agriculture 13(5):1073

Chen P, Zhu G, Kim HJ, Brown PB, Huang JY (2020) Comparative life cycle assessment of aquaponics and hydroponics in the Midwestern United States. J Clean Prod 275:122888

Chen C, Tang T, Shi Q, Zhou Z, Fan J (2022) The potential and challenge of microalgae as promising future food sources. Trends Food Sci Technol 126:99–112

Cisternas I, Velásquez I, Caro A, Rodríguez A (2020) Systematic literature review of implementations of precision agriculture. Comput Electron Agric 176:105626

Dolganyuk V, Belova D, Babich O, Prosekov A, Ivanova S, Katserov D, Patyukov N, Sukhikh S (2020) Microalgae: a promising source of valuable bioproducts. Biomolecules 10(8):1153

Edo M, Entling MH, Rösch V (2024) Agroforestry supports high bird diversity in European farmland. Agron Sustain Dev 44:1

Fabi C, Cachia F, Conforti P, English A, Moncayo JR (2021) Improving data on food losses and waste: from theory to practice. Food Policy 98:101934

Finger R, Swinton SM, El Benni N, Walter A (2019) Precision farming at the nexus of agricultural production and the environment. Ann Rev Resour Econ 11(1):313–335

Forejt M, Syrbe RU (2019) The current status of orchard meadows in Central Europe: multi-source area estimation in Saxony (Germany) and the Czech Republic. Morav Geogr Rep 27(4):217–228

Fussy A, Papenbrock J (2022) An overview of soil and soilless cultivation techniques—chances, challenges and the neglected question of sustainability. Plants 11(9):1153

Hashempour-Baltork F, Khosravi-Darani K, Hosseini H, Farshi P, Reihani SFS (2020) Mycoproteins as safe meat substitutes. J Clean Prod 253:119958

Jan S, Rashid Z, Ahngar TA, Iqbal S, Naikoo MA, Majeed S,... Nazir I (2020) Hydroponics–a review. Int J Curr Microbiol Appl Sci 9(8):1779–1787

Joshi P (2020) A Review On High Density Planting In Fruit Crops. Planta 1:84–98

Jurić M, Bandić LM, Carullo D, Jurić S (2024) Technological advancements in edible coatings: emerging trends and applications in sustainable food preservation. Food Biosci 58:103835

Jürkenbeck K, Heumann A, Spiller A (2019) Sustainability matters: consumer acceptance of different vertical farming systems. Sustainability 11(15):4052

Khan R, Brishti FH, Arulrajah B, Goh YM, Abd Rahim MH, Karim R, Hajar-Azhari S, Kin Kit S, Anwar F, Saari N (2024) Mycoprotein as a meat substitute: production, functional properties, and current challenges-a review. Int J Food Sci Technol 59(1):522–544

Khatri L, Kunwar A, Bist DR (2024) Hydroponics: advantages and challenges in soilless farming. Big Data Agric (BDA) 6:81–88

Kinoshita T, Yamazaki H, Inamoto K, Yamazaki H (2016) Analysis of yield components and dry matter production in a simplified soilless tomato culture system by using controlled-release fertilizers during summer–winter greenhouse production. Sci Hortic 202:17–24

Kumar K, Gambhir G, Dass A, Tripathi AK, Singh A, Jha AK, Yadava P, Choudhary M, Rakshit S (2020) Genetically modified crops: current status and future prospects. Planta 251(4):91

Kumar P, Sharma N, Sharma S, Mehta N, Verma AK, Chemmalar S, Sazili AQ (2021) In-vitro meat: a promising solution for sustainability of meat sector. J Anim Sci Technol 63(4):693

Kwon CT, Heo J, Lemmon ZH, Capua Y, Hutton SF, Van Eck J, Park SJ, Lippman ZB (2020) Rapid customization of Solanaceae fruit crops for urban agriculture. Nat Biotechnol 38(2):182–188

Kyriakopoulou K, Dekkers B, van der Goot AJ (2019) Plant-based meat analogues. In: Sustainable meat production and processing. Academic, pp 103–126

Laila A, Topakas N, Farr E, Haines J, Ma DW, Newton G, Buchholz AC (2021) Barriers and facilitators of household provision of dairy and plant-based dairy alternatives in families with preschool-age children. Public Health Nutr 24(17):5673–5685

Lal R, Reicosky DC, Hanson JD (2007) Evolution of the plow over 10,000 years and the rationale for no-till farming. Soil Tillage Res 93(1):1–12

Lennard W, Goddek S (2019) Aquaponics: the basics. In: Aquaponics food production systems. Springer, Cham, p 113

Lipper L, Thornton P, Campbell BM, Baedeker T, Braimoh A, Bwalya M, Caron P, Cattaneo A, Garrity D, Henry K, Hottle R (2014) Climate-smart agriculture for food security. Nat Clim Chang 4(12):1068–1072

Lucakova S, Branyikova I, Hayes M (2022) Microalgal proteins and bioactives for food, feed, and other applications. Appl Sci 12(9):4402

Martínez-Romero E, Aguirre-Noyola JL, Taco-Taype N, Martínez-Romero J, Zuniga-Davila D (2020) Plant microbiota modified by plant domestication. Syst Appl Microbiol 43(5):126106

Marvin HJ, Bouzembrak Y, Van der Fels-Klerx HJ, Kempenaar C, Veerkamp R, Chauhan A, Stroosnijder S, Top J, Simsek-Senel G, Vrolijk H, Knibbe WJ (2022) Digitalisation and artificial intelligence for sustainable food systems. Trends Food Sci Technol 120:344–348

Mellor C, Embling R, Neilson L, Randall T, Wakeham C, Lee MD, Wilkinson LL (2022) Consumer knowledge and acceptance of "algae" as a protein alternative: a UK-based qualitative study. Foods 11(12):1703

Michler JD, Baylis K, Arends-Kuenning M, Mazvimavi K (2019) Conservation agriculture and climate resilience. J Environ Econ Manag 93:148–169

Mozaffari A, Nouman M (2021) Exploring the drivers and moderators of business model innovation: a case study of the dairy industry. http://lup.lub.lu.se/student-papers/record/9055613

Musicus AA, Wang DD, Janiszewski M, Eshel G, Blondin SA, Willett W, Stampfer MJ (2022) Health and environmental impacts of plant-rich dietary patterns: a US prospective cohort study. Lancet Planet Heath 6(11):e892–e900

Noack F, Engist D, Gantois J, Gaur V, Hyjazie BF, Larsen A, M'Gonigle LK, Missirian A, Qaim M, Sargent RD, Souza-Rodrigues E (2024) Environmental impacts of genetically modified crops. Science 385(6712):eado9340

Ogunyemi SO, Abdallah Y, Zhang M, Fouad H, Hong X, Ibrahim E, Masum MMI, Hossain A, Mo J, Li B (2019) Green synthesis of zinc oxide nanoparticles using different plant extracts and their antibacterial activity against Xanthomonas oryzae pv. oryzae. Artif Cells Nanomed Biotechnol 47(1):341–352

Patnaik S, Sen S, Mahmoud MS (2020) Smart village technology. In: Modeling and optimization in science and technologies, vol 17, Springer Nature, pp 181–189

Pawlak K, Kołodziejczak M (2020) The role of agriculture in ensuring food security in developing countries: considerations in the context of the problem of sustainable food production. Sustainability 12(13):5488

Pinstrup-Andersen P (2018) Is it time to take vertical indoor farming seriously? Glob Food Sec 17:233–235

Pramanik P, Krishnan P, Maity A, Mridha N, Mukherjee A, Rai V (2020) Application of nanotechnology in agriculture. In: Environmental nanotechnology, vol 4. Springer International Publishing, Cham, pp 317–348

Pulizzi F (2019) Nano in the future of crops. Nat Nanotechnol 14(6):507–507

Ragaveena S, Shirly Edward A, Surendran U (2021) Smart controlled environment agriculture methods: a holistic review. Rev Environ Sci Biotechnol 20(4):887–913

Raj EFI, Appadurai M, Athiappan K (2022) Precision farming in modern agriculture. In: Smart agriculture automation using advanced technologies: data analytics and machine learning, cloud architecture, automation and IoT. Singapore, Springer Singapore, pp 61–87

Rakocy JE (2012) Aquaponics—integrating fish and plant culture. In: Aquaculture production systems, Wiley, pp 344–386

Rizwan M, Mujtaba G, Memon SA, Lee K, Rashid N (2018) Exploring the potential of microalgae for new biotechnology applications and beyond: a review. Renew Sust Energ Rev 92:394–404

Rockström J, Edenhofer O, Gaertner J, DeClerck F (2020) Planet-proofing the global food system. Nat Food 1(1):3–5

Rodriguez-Ortega WM, Martinez V, Rivero RM, Camara-Zapata JM, Mestre T, Garcia-Sanchez F (2017) Use of a smart irrigation system to study the effects of irrigation management on the agronomic and physiological responses of tomato plants grown under different temperatures regimes. Agric Water Manag 183:158–168

Roy B, Hagappa A, Ramalingam YD, Mahalingam N (2021) A review on lab-grown meat: advantages and disadvantages. Quest Int J Med Health Sci 4(1):19–24

Sadar J, Chyon G (2020) Designing with fog. In: Proceedings of the 4th international congress on ambiances, alloaesthesia: senses, inventions, worlds, vol 1. Réseau International Ambiances, p 168

Samtiya M, Aluko RE, Dhewa T (2020) Plant food anti-nutritional factors and their reduction strategies: an overview. Food Prod Process Nutr 2(1):6

Sánchez-Bayo F, Wyckhuys KA (2019) Worldwide decline of the entomofauna: a review of its drivers. Biol Conserv 232:8–27

Sandhu R, Chaudhary N, Shams R, Dash KK (2025) Genetically modified crops and sustainable development: navigating challenges and opportunities. Food Sci Biotechnol 34(2):307–323

Sardare MD, Admane SV (2013) A review on plant without soil-hydroponics. Int J Res Eng Technol 2(3):299–304

Singh B (2023) New systems of vegetable production: protected cultivation, hydroponics, aeroponics, vertical, organic, microgreens. In: Vegetables for nutrition and entrepreneurship. Singapore, Springer Nature Singapore, pp 31–56

Smetana S, Sandmann M, Rohn S, Pleissner D, Heinz V (2017) Autotrophic and heterotrophic microalgae and cyanobacteria cultivation for food and feed: life cycle assessment. Bioresour Technol 245:162–170

Srivastav AL, Dhyani R, Ranjan M, Madhav S, Sillanpää M (2021) Climate-resilient strategies for sustainable management of water resources and agriculture. Environ Sci Pollut Res 28(31):41576–41595

Suresh A, Harsha S, Surya K, Laha D, Gaba D, Bhardwaj S, Teja V, Bhambri S, Bhambri S (2017) Innovative human mars mission with vertical farming. In: Mars society convention, ASWATH pp 25–29

Szczebyło A, Rejman K, Halicka E, Laskowski W (2020) Towards more sustainable diets—Attitudes, opportunities and barriers to fostering pulse consumption in Polish cities. Nutrients 12(6):1589

Tachie C, Nwachukwu ID, Aryee AN (2023) Trends and innovations in the formulation of plant-based foods. Food Prod Process Nutr 5(1):16

Taha MF, ElMasry G, Gouda M, Zhou L, Liang N, Abdalla A, Rousseau D, Qiu Z (2022) Recent advances of smart systems and internet of things (iot) for aquaponics automation: a comprehensive overview. Chem 10(8):303

Tangyu M, Muller J, Bolten CJ, Wittmann C (2019) Fermentation of plant-based milk alternatives for improved flavour and nutritional value. Appl Microbiol Biotechnol 103(23):9263–9275

Tilman D, Balzer C, Hill J, Befort BL (2011) Global food demand and the sustainable intensification of agriculture. Proc Natl Acad Sci 108(50):20260–20264

Tso R, Forde CG (2021) Unintended consequences: nutritional impact and potential pitfalls of switching from animal-to plant-based foods. Nutrients 13(8):2527

Tuomisto HL, Teixeira de Mattos MJ (2011) Environmental impacts of cultured meat production. Environ Sci Technol 45(14):6117–6123

Uddin MR, Suliaman MF (2021) Energy efficient smart indoor fogponics farming system. In: IOP conference series: earth and environmental science, vol 673, No. 1. IOP Publishing, p 012012

Van Gerrewey T, Boon N, Geelen D (2021) Vertical farming: the only way is up? Agronomy 12(1):2

Venkatesh S, Victor Ebenezer RD, Vignesh AC, Vishnu A, Sree Lekha S (2020) A study on development of crops by fogponic system using coco coir. Int Res J Eng Technol 7(4):6043–6047

Vishnoi S, Goel RK (2024) Climate smart agriculture for sustainable productivity and healthy landscapes. Environ Sci Pol 151:103600

Walls H, Baker P, Chirwa E, Hawkins B (2019) Food security, food safety & healthy nutrition: are they compatible? Glob Food Sec 21:69–71

Xi L, Zhang M, Zhang L, Lew TT, Lam YM (2022) Novel materials for urban farming. Adv Mater 34(25):2105009

Yang S, Xu H, Chen JH, Liu B, Cheng KW (2024) Microalgae as a sustainable protein source: key issues related to their production, application, and the way forward. Food Bioprocess Technol 17(8):1–33

# Chapter 5
# Climate Resilience Through Plant Diversity

**Abstract** Most people concur that the biggest threat to life on Earth is the rapid change in the climate. There has been no evaluation of the possible effects of climate change on the diverse range of plant species. Depending on their natural distribution, the predicted changes, their capacity for adaptation and resilience, their ability to migrate and disperse, the ecology and nature of their new bioclimatic envelopes, and their capacity to endure in and colonize new areas, the effects of climate change on plants will differ. The environment, potential management approaches, occurrence of pollinators and dispersal agents also serves as significant factors. Hence, the present chapter highlights the role of plant diversification in enhancing climate resilience addressing various stressors along with indigenous knowledge to enhance their productivity. It also addresses ecosystem services and the importance of plants to improve soil health by mitigating the adverse effects posed by environment.

**Keywords** Climate change · Ecosystem services · Indigenous knowledge · Plant diversity · Resilience

## 5.1 Introduction

In order to guarantee future food security, it is imperative to boost agricultural yields in a sustainable way while simultaneously rendering crops more resilient to climate change. Extreme heat and unpredictable rainfall brought on by climate change will make it more difficult for agriculture to supply the growing world population's food needs. Quick developments in understanding and applied crop design could offer answers based on cutting-edge techniques, such as gene editing, genomics, soil microbiome modification, and crop management techniques (Diffenbaugh et al. 2017). Assessing current knowledge and taking into account how cooperative efforts that stimulate and promote better interactions within disciplines may be necessary to accomplish agri-food system targets are necessary given the size of the climate change challenge. Extreme weather events are expected to become more frequent,



N. Patel et al., *The Future of Food: Exploring Edible Plants for a Sustainable Diet*,
https://doi.org/10.1007/978-3-032-15027-1_5

intense, longer-lasting, and have larger spatial dimensions due to climate change, which poses a serious danger to the production of all major crops (Rogelj et al. 2012). For instance, the yields of cereal crops like wheat and maize are negatively impacted by an increase in heatwaves, droughts, and floods; during drought circumstances, these yields can drop by as much as 40% (Daryanto et al. 2016).

The crop, range and intensity of the stress, and the stage of crop growth all influence the extent to which yield is lost. Since heat waves and droughts reduce yields in most crop periods, the loss in yield is anticipated to persist, especially for reproduction and grain development/filling. Additionally, the consequences of temperature variations are exacerbated by the fact that 40% of durum wheat, 55% of barley, 40% of alfalfa and 30% of grass are currently grown in drought-prone areas (Khatun et al. 2021). To ensure food security, worldwide agricultural production needs to rise by 25% over the next 10 years, despite the limitations imposed by climate change. It is anticipated that as the world's population grows and severe weather patterns become increasingly prevalent, food insecurity might deteriorate. The amount of arable land must grow from 1.0 to 1.5 billion hectares in order to provide food security for the 10 billion people that are expected to inhabit the planet by 2050, while the amount of water available for agriculture will decrease (Deryng et al. 2014). Crop breeding must undergo paradigm-shifting advancements to guarantee that knowledge gains on climate resilience are successfully converted into improved crops. Furthermore, in order to benefit from contemporary genetic advancements, they must be made efficiently, with the least possible negative effects on the environment, within financial limits, with social acceptance, and under unified administration (Asseng et al. 2015).

In an era marked by shifting human activities, catastrophic events and climate change, the diversity and abundance of plant species within communities are important determinants of the ecosystem's resilience (Fig. 5.1). Located at the base of the ecological pyramid, plants are essential species that give many other species resources, food, and shelter (Çekiç, et al. 2012).

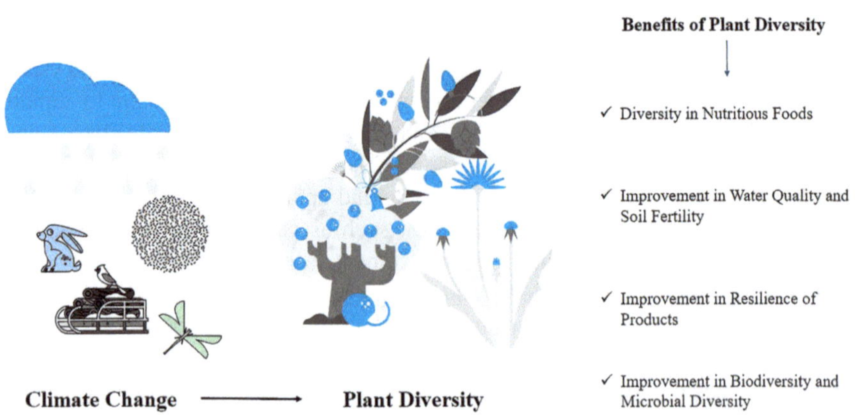


**Fig. 5.1** Benefits of plant diversity on climate change

Plants are essential to mitigate the consequences of climate change. In order to preserve the intricate equilibrium required for life, these silent guardians of nature serve as the planet's lungs, taking in carbon dioxide from the atmosphere and releasing oxygen (Sheehy et al. 2005). In addition to their aesthetic and ecological significance, plants are vital partners in the worldwide battle against climate change. Trees, especially tropical rainforests like the Amazon, are sometimes referred to as carbon sinks due to the enormous amounts of carbon they trap in their substrates and biomass. Additionally, mangroves, wetlands, and grasslands are essential for storing carbon. In their wet soils, mangroves, for example, are incredibly effective at absorbing and storing approximately four times as much carbon as terrestrial forests. The amount of greenhouse gasses in the atmosphere can be considerably decreased by maintaining and growing these ecosystems (Hawkins et al. 2008).

Sustainable farming methods, like agroforestry, for instance, combine trees with crops and livestock in order to improve soil health and lessen the need for artificial fertilizers. Carbon is efficiently stored in healthy soils that are high in organic matter, lowering its release into the atmosphere. Furthermore, there are numerous ways in which plants could eventually replace the place of fossil fuels. Plant-based biomass is a renewable energy source that can replace natural gas, coal, and oil (Engler et al. 2009). Compared to conventional fuels, biofuels made from crops like corn and sugarcane release fewer pollutants, enabling a cleaner energy future. Additionally, diversified plant ecosystems increase resistance to the consequences of climate change, including droughts and floods.

In addition to stabilizing soils and minimizing erosion, forests and grasslands enhance water retention, which lessens the intensity of floods. Although lowering carbon dioxide is the main goal to fight against climate change, plants also enhance air quality by absorbing other pollutants like nitrogen oxides, sulfur dioxide and ammonia (Rozema et al. 2007). In addition to storing carbon, urban green spaces such as parks and rooftop gardens also filter particulate matter, providing city inhabitants with healthier air.

Considering their enormous potential, plants are threatened in numerous capacities resulting in it challenging for them to confront climate change. Forest cover is declining at startling rates due to deforestation caused by logging, agriculture, and urbanization. Billions of tons of $CO_2$ were released into the atmosphere in 2023 alone as tropical forests lost an area in Switzerland. Plant ecosystems are under stress from rising temperatures, changing rainfall patterns and extreme weather events, which lowers plant productivity and growth. Native vegetation is further threatened by invasive species and pests, which are frequently spreading as a result of warmer conditions (Akanwa et al. 2019). Despite strong evidence, there is still a dearth of studies explicitly examining the role that species diversity plays in ensuring food security and climate resilience. Bridging this gap is essential to develop workable plans to support food security and diversity in various agricultural settings.

## 5.2 Plants Amidst Overlapping Challenges

Numerous stressors, whether biotic or abiotic such as soil-associated or climate-driven, human-made, or both can continue to affect plants. Plants frequently experience multiple stressors at once, and these stressors may interact with one another. This interaction may be additive, antagonistic or synergistic in nature. Synergy happens when the combined impact of several stressors is higher than the average of the impacts of each stressor alone (Zandalinas et al. 2021). Conversely, antagonistic relationship occur when the stressors' combined impact is less than the total of their separate effects. The concept of additive interaction suggests that the total effect of numerous stressors is equivalent to the sum of their individual effects.

Stress interactions vary depending on the plant they impact as well as the conditions that cause them. One species may be negatively impacted by a stressor combination whereas another may be positively impacted. Plant homeostasis is disrupted by stress, which impacts physiological and metabolic functions. Throughout their lives, plants can quickly adjust to changes in their environment by establishing a new state of homeostasis. Acclimatization is the term used to describe this adaptive process (Zandalinas and Mittler 2022). Over time, genetic changes in the population brought on by modifications in the genome can also cause plants to adapt to a new environmental pattern; this process is called adaptation.

A number studies concentrate on how plants react to a single stressor or to straightforward stress combinations that involve two or, at most, three distinct stressors happening at the same time. However, plants frequently face a variety of stressors in the natural world. A multivariate stress compilation is the result of three or more stresses acting on a plant at the same time. Soil drought combined with high air temperatures is the most frequently researched basic combination of stressors (Kleine et al. 2021). Stomatal closure is triggered by signals sent from the roots to the shoot during water deficiencies. The evacuation of water from the tissues can be prevented by stomatal closure. Conversely, warm weather encourages stomatal opening, which enables plants to use transpiration to cool themselves. When both events occur at the same time, stomatal closure frequently occurs in the foliage but not in the blossoms, which stay open. This safeguards the reproductive process and stops tissue from overheating. This adaptation technique is known as "differential transpiration." One benefit of stomatal closure of leaves is that it reduces the spread of pathogens to the leaf, which is especially vulnerable to plants in adverse environments (Sinha et al. 2022). Elevated $CO_2$ concentrations in the natural environment are now present in conjunction with drought and heat stress. In C3 photosynthesis crops, e[CO2] results in decreased stomatal conductance and higher photosynthetic carbon absorption. Because of stomatal closure, leaves' intercellular $CO_2$ content is not stimulated, which is essential for better carbon absorption at elevated $CO_2$. When soil water shortage and elevated $CO_2$ are combined, legume nodulation is disrupted and tissue nitrogen levels drop. Under ideal growth conditions, C4 crop photosynthesis is typically not directly stimulated by rising atmospheric $CO_2$ concentrations. Nevertheless, elevated $CO_2$ might lessen the detrimental effects of heatwaves and increase crop production in regions with limited water supplies (Ottman et al. 2001; Fitzgerald et al. 2016).

The same or distinct portions of a plant may perceive distinct stresses that are affecting it at the same time. For example, the roots may sense heavy metals in the soil, the leaves may sense powdery mildew, and the entire plant may sense heat stress. Different sensors and signal transduction networks are used by plants to respond to environmental stimuli. Every stress sets off unique physiological and metabolic processes as well as systemic signaling and acclimatization responses (Gray et al. 2016). Plants are able to incorporate systemic messages, and how they react will rely on how they perceive the variables that are generating these signals.

The overall condition of the soil determines the high production and nutritional value of a large number of farmed plants. Researchers used a variety of combinations of drought, low nitrogen, temperature, microplastics, glyphosate, antibiotics, fungicides, copper, salt, and insecticides to investigate the effects of stress factors linked to climate change and human activities in soils. Negative impacts on soil characteristics and its microbiota were shown to be exacerbated by an upsurge in the amount of concurrent factors that make up a multifactorial stress combination. The soil is negatively impacted by many of the same causes that negatively influence plants. Therefore, another element that stresses crops is the degradation of soil qualities (Ma et al. 2022).

## 5.3 Plant Diversity Driven Soil Health

Soil health is greatly impacted by diversifying plants, especially when it comes to the supply of vital ecosystem services. Different varieties of plants exhibit distinct levels of tolerance and resilience in various climatic conditions. Through their interactions, intricate networks of material cycling and supply chains can be established within ecosystems, improving resilience and sustainability. Additionally, plant diversity lessens reliance on chemical fertilizers and pesticides, increases soil nutrient use efficiency, and minimizes the negative environmental effects on agriculture (Rebouh et al. 2023). The objectives of environmentally friendly agriculture can be advanced by increasing agricultural land production and quality through the adoption of a range of crops and their mutually beneficial effects. Furthermore, a soil health and ecological well-being is fostered by growing a variety of vegetation, which improves the porosity and fertility of soil, optimize the structure of soil microbiota and raise the amount of organic carbon in the soil (Toor et al. 2021).

By increasing resistance and flexibility, the emergence of diverse crop species can lower the prevalence of pests and diseases. Plant diversification based on ecological niches promotes efficient use of soil nutrients, reduces soil degradation and maintains soil fertility and profitability. Because various species perform unique roles and activities within the ecosystem while building networks of interactions in their shared environment, a higher level of species variety is linked to increased production (Ferris and Tuomisto 2015). The issue of mineral depletion that results from growing the same crop for an extended period of time on a single piece of land can be mitigated by different crop varieties using different soil nutrients. Plant growth and development can be aided by the complementing impacts of various crops, which can maximize the use of soil resources and improve the efficiency of nutrient use (Chen et al. 2022).

In order to encourage diversity, crop rotation and intercropping are crucial management techniques. The quality and productivity of soil can be greatly increased by growing multiple crops at once or rotating them. Crop rotation reduces the buildup and spread of soil-borne illnesses and pests that come with growing a single crop continuously. Additionally, different planting techniques can change the soil's microbial makeup, which helps control diseases and pests and lessens the need for pesticides (Chahal et al. 2021; Das et al. 2022). It also improves the soil's ability to use nutrients efficiently. The fertilizer demands of various planting techniques vary, which can reduce nutrient waste and increase nutrient recycling rates. The variety of soil microorganisms can be increased with the help of intercropping by the synergistic interactions between different crops. Exudates from the roots of numerous plants can improve the organic matter content, strengthen the soil's structure, and encourage the growth and multiplication of soil microorganisms (Upadhyay et al. 2022). Furthermore, the differences in development cycles and traits of crops can maximize the use of soil resources and enhance the overall condition of the soil. It is feasible to improve the soil microbes, minimize the prevalence of soil pests and diseases, and increase nutrient intake efficiency by carefully planning crop diversification and intercropping systems. All of these factors contribute to the long-term sustainability of agricultural production (Qiao et al. 2022).

## 5.4 Perennial Landscapes For Mitigating Climate Change

Compared to annual crops, perennial systems have the potential to improve ecosystem services from agricultural land. Investigation conducted across natural perennial grassland environments provides a large portion of this potential. The advantages of perennials in relation to of global warming are obvious and include qualities that reduce negative environmental effects through climate regulation, decreased nutrient loss to waterways, increased biodiversity, and reduced greenhouse gas emissions (e.g., reduced tillage, more efficient N uptake, and dual-use potential) (Ricciardi et al. 2018). Through its resilience to extreme climate events and stability in production over time, perennial crops also aid in climate change resilience.

In order to sustain production of food and ecological function in the face of global climate change, farmers and politicians can establish targets by developing credible projections of future temperatures. Temperature variations, temperature patterns like freeze/thaw cycles, diminishing precipitation, extreme weather events like floods, storms, and droughts, and changes in infectious disease and pest distributions must all be accurately predicted in these forecasts (Asbjornsen et al. 2014). Genetics (e.g., generating new crops, introducing novel characteristics, gene engineering, breeding for resilience through genetic variability) and management (e.g., shifting crop ecoregions, altering agronomic and horticultural practices) are the two main categories into which adaptation strategies fall. These strategies could increase agricultural system adaptability to changing climates. In order for perennial systems to fulfill the various goals of farmers, society, and markets, it is necessary to enhance horticultural practices, improve agronomics, improve germplasm, and comprehend environmental variation (Ryan et al. 2018). This is because each of these factors has distinct time requirements that allow for their adoption on the landscape. The selection cycles of perennial crops are usually longer, making it more challenging to sustain genetic gain at the required pace of change. Considering that certain systems are short-lived (2–3 years) and others are in production for comparatively longer periods of time (20–30 years), it is necessary to have more precise strategies and selection criteria for the various levels of perenniality (Leisner 2020).

Agricultural productivity can be safeguarded against the effects of global warming by having well-functioning soil and water resources. Because of their vast root systems, lengthy growing seasons, and lack of yearly tillage or soil disturbance for ordinary production, perennial plants have been demonstrated to improve soil functioning and water control. Therefore, it may be possible to restore soil quality by cultivating perennial crops on marginal soils, halting desertification and opening up additional acreage for provisioning advantages (Fatichi et al. 2020). Enhancing soil structure, for instance, can lower surface-level runoff, enhance irrigation water uptake, and raise penetration rates in soils susceptible to superficial compression. Raising the amount of organic matter in the soil can improve crop water uptake and its ability to hold water. The benefits of meadows and forage crops to ecosystems are widely recognized, and they come in a wide variety of cropping systems. One example of how perennial crops can enhance soil quality and water control is the perennial grain Kernza. Perennials are a potent tool for climate adaptation in part because of their capacity to deliver these ecosystem advantages (Audu et al. 2022).

Additionally, because perennials have an extended growing season than annuals, they can absorb and use water that is accessible outside of the usual annual crop producing season. The loss of nutrients to both ground and surface waters can be minimized by using water that annual crops cannot use. According study findings, growing intermediate wheatgrass for grain can lessen NO3 leaching to groundwater, safeguarding rural drinking water supplies and providing farmers with a profit. Similarly, it was demonstrated that nitrate N losses and subsurface field outflow volume were 30–50 times greater under annual crops than under perennial crops (Jungers et al. 2019; Reilly et al. 2022).

## 5.5 Ecosystem Services Provided by Plant Diversity

Biodiversity includes all forms of life on Earth, including species, genetic, and ecological diversity. This concept recognizes the intricate web of relationships that exist between various species and their environments. However, there are various benefits that ecosystems provide to humans, including ecosystem/ecological services (Mace et al. 2012). These benefits fall into four main categories: provisioning services, which include things like food, water, and raw materials; regulating services, which include things like climate regulation and pest control; cultural services, which include things like aesthetic and recreational values; and supporting services, which include things like nutrient cycling and pedogenesis (Loreau et al. 2001; Tilman et al. 2014). Plants, biological diversity and ecosystem services are all closely related. In farming systems, diverse plant and animal species are better suited in enhancing ecosystem functionality and resilience. Furthermore, biological diversity contributes to the sustainability of cultivation practices by enhancing ecological services important for the food production. For example, many crops depend on pollinators, such as bees and butterflies, to reproduce successfully. Similarly, nutrient cycling, which ensures that plants receive the nutrients they require to flourish, depends on soil microbes. To develop agricultural practices that are not only productive but also sustainable and resistant in enduring environmental issues, it is essential to understand and promote the intricate relationships between ecosystem services and biodiversity (Garibaldi et al. 2013).

One important ecological service affecting plant yield and affordability of food is pollination. The diversification process leads to healthy populations of pollinators such as birds, honeybees, butterflies and or other insects (Potts et al. 2006). Most of the products for human sustenance comes from flowering plants and approximately 30–35% of food crops and 70–75% of flowering plants requires pollination. This includes diverse plants such as fruits, vegetables and nuts. The productivity of crop is enhanced by pollinators for instance, as observed in berries and almonds. Diverse biological system can also serve as a buffer against environmental changes that threaten pollinator populations, such as habitat degradation and climate fluctuations (Potts et al. 2010).

## 5.6 Climate Resilient Plants: The Simple Difference in Changing World

*Secale cereale* L (Rye; Poaceae family) is grown extensively in Central and Eastern Europe. This grain is resistant to a number of diseases and unfavorable environmental factors, and it shows a high tolerance to location, forecrop and fertilization. *Secale cereale* var. multicaule, or primitive rye, has gained popularity recently, especially for grain production in hilly areas because of its ability to withstand cold and bad soil conditions (Dzienis 2018). When sown in the spring, primitive rye is a biennial plant that can be harvested for seed the next year and biomass for silage the same year. As an alternative, it is grown similarly to winter rye when sowed in the fall. Weeds are effectively suppressed by the plant's exceptional tillering ability. Additionally, it can be applied to agricultural soil regeneration (Vejražka et al. 2009).

Notably, *Sorghum bicolor* L.; Poaceae family is one of the most significant carbohydrate-rich crops in the world, after rice, wheat, corn, and barley. In addition to being an important fodder crop in many nations across the world, sorghum is a staple diet for people in Asia and Africa. Because of its ability to withstand soil drought and high air temperatures, it is a recommended option in semi-arid climates and locations with erratic rainfall. It can also be grown on marginal farmland with low inputs (Chadalavada et al. 2021; Hossain et al. 2022).

The plant known as pearl millet (*Pennisetum glaucum* L.; Poaceae family) is very popular in Sub-Saharan Africa. Its ability to withstand extended periods of high temperatures and soil water deficiencies is a noteworthy benefit. During the reproductive stage, it can withstand drought and intense heat and thrives in regions with higher salinity (Satyavathi et al. 2021). Anticipated as a future crop, amaranth (*Amaranthus* spp. L.; Amaranthaceae family) is a pseudocereal that has been known in Central and South America since ancient times. *A. cruentus* L., *A. caudatus* L., and *A. hypochondriacus* L. are the three species that are farmed. Because of its great genetic variety and wide range of phenotypic flexibility, amaranth is a plant that thrives on soils with low nutrient levels. It can withstand high air temperatures, excessive soil salinity, and a lack of soil water (Reyes-Rosales et al. 2023). Quinoa (*Chenopodium quinoa* Willd.; Amaranthaceae family) is another pseudocereal that came from the same area as amaranth and has significant climate change resilience. Although it is from South America, quinoa can be cultivated in low-input, rocky farms. It has deep roots systems, great water-use efficiency, and tolerance to cold temperatures, salt, and water scarcity (Heinz et al. 2024).

The ancient European plant Camelina (*Camelina sativa* L.; Brassicaceae family) is gaining popularity due to its adaptability to a variety of growing environments. With the exception of thick clay and organic soil, *Camelina* may grow in a variety of climates and soil types. When other crops fail, it still produces successfully. It produces low-emission biodiesel fuel and is used as an oilseed crop in the food, feed, and biomedical industries. *Camelina* is a low-input crop that can withstand soil drought, little rainfall, and a brief vegetation period since it is suited to chilly temperate semi-arid areas (Waraich et al. 2013).

## 5.7 Indigenous Knowledge Towards Climate Resilience

There is no single canonical definition for Indigenous people, Indigenous knowledge, and Indigenous food systems, and the name "Indigenous" is ambiguous. The word "Indigenous" frequently describes an entity's native, traditional, or inherited characteristics in a given geographic area. Indigenous knowledge is defined as "any implementation of those traditions as living knowledge to improve present and future circumstances," and an Indigenous person is defined as "a person belonging to a community preserving memories of life lived sustainably on a land-base, as part associated with that land-base." (Yunkaporta 2019) Indigenous peoples maintain a strong spiritual bond with the natural world. Indigenous Peoples' belief systems, identities, and knowledge are based on their close ties to locations of special significance. They have an impact on how Aboriginal individuals react to changes in the climate (Ford et al. 2020). The food systems of Indigenous Peoples sustain biodiversity while giving their communities access to food, livelihoods, sustenance, and products. Their strong ties to sustainable maintenance of marine and terrestrial resources can contribute to ecosystem resilience. Based on their historical, cultural, and spiritual ties to the land they have inhabited for generations, Indigenous Peoples have maintained, conserved, and safeguarded their environment through customary practices and duties. These activities form the basis of their stewardship (Sanchez et al. 2023).

Indigenous Peoples' and local communities' knowledge and values offer superior methods for dealing with environmental and social challenges and adjusting to constant change. This is particularly true given that 50% of our current caloric intake comes from wheat, paddy and maize, and that crop diversification is declining globally. Around 300 million people rely on woods for their sustenance, and an estimated one million people use wild foods in their diet worldwide. Advocating for the use of Indigenous Peoples' knowledge and customs begins with acknowledging their worth (Brondízio et al. 2021). Though their knowledge systems have co-evolved with ecosystems and they retain over 80% of the world's biodiversity, Indigenous Peoples' food systems guarantee ecological and socioeconomic sustainability, but they continue to be sidelined in research and politics. Traditional improvements in agriculture frequently overlook how adaptive Indigenous knowledge and techniques are. The argument over local versus global food systems needs to be reframed from a resilience perspective. Recognizing that Indigenous people have lessons to teach about sustainability is crucial. For more than 60,000 years, Australian biocultural knowledge, land use, and ancestral methods have formed the ecosystems of the country's unique flora and animals. These techniques are now acknowledged and incorporated into conservation initiatives. The importance of Indigenous knowledge in fostering ecological resilience, transformational stewardship techniques, and cultural revitalization has also gained more recognition. Traditional science instruction can be improved and supplemented by indigenous knowledge (Wood et al. 2023).

Our existing food systems have four "aching points." These include food insecurity and trade, inequality and governance as significant vulnerabilities, environmental and nutritional deterioration, and food system illiteracy as a critical issue that has to be addressed. Cultural and nutritional poverty has been brought about by the simplification of landscapes, particularly the "erosion" of Indigenous and traditional food plants. It is obvious that our current food systems urgently need to be improved in the area of biodiversity loss with food insecurity and nutritional repercussions. Important starting points for learning about food from aboriginal peoples' knowledge base include the provision of sustainable, healthful foods and the usage of low- and renewable energy sources. Indigenous peoples have the potential to play a greater role in facilitating the lessons learned from their sustainable farming practices in order to improve environmental protection and the sustainability of global food systems (Vijayan et al. 2022).

## 5.8 Challenges and Potential Strategies to Improve Plant Resilience to Climate Change

Genome-Wide Association Studies, other genetic mapping and sequencing tools, and sophisticated phenotyping methodologies are currently being used to investigate the genetic variation of plants at the single nucleotide level. The use of large-scale sequencing, metabolomics, machine learning tools, and CRISPR-Cas technologies has also significantly advanced our understanding of the temporal and geographical regulation of gene expression and metabolic pathways (Rivero et al. 2022). However, an understanding of the procedures that control plant productivity, growth, and yield in field settings and stress combinations is crucial for the effective implementation of these new technologies. Understanding the metabolic, physiological, and developmental mechanisms that plants use to react to and communicate with their changing environment, pathogens, and pollinators will be important. A "balanced" genetic makeup is necessary for new and improved plant types and cultivars in order to reduce losses brought on by individual environmental challenges, pests and other quickly shifting environmental variables (Kissoudis et al. 2016).

Among the key goals include improving the efficiency of photosynthesis, modifying stomatal control, establishing a balanced and effective use of water and nutrients, and promoting plant interactions with advantageous microbes. Considering the anticipated adverse climatic circumstances brought on by climate change, some potential strategies to boost plant productivity and yield includes, Increasing resilience to stressor combinations while paying particular attention to various climate scenarios that involve crops being exposed to heat in conjunction with drought, salinity, flooding, elevated CO2, and/or disease threat. Better knowledge of the physiological, metabolic, and signaling processes that underlie crop responses to diverse stress combinations as well as the beneficial and detrimental interactions between various stresses will be necessary for the position to succeed (Mondal et al. 2023). Achieving this challenging objective requires constant collaboration and candid multidisciplinary communication between many researchers enhancing the processes of seed filling and plant reproduction in combined stress situations. Enhancing the ability of flowers to withstand heat and desiccation as well as the distribution of photo assimilates from leaves to flowers could help achieve this goal. It is especially important to focus on stress combination scenarios that involve extreme temperatures because heat has been shown to negatively impact blooming and reproduction. In order to help farmers mitigate certain aspects of climate change in real time on the ground, precision farming methods should be developed, introduced, and improved. This objective includes the development of novel chemical use and irrigation techniques, enhanced robotics and drone technologies, and better imaging and mapping technologies (Kopeć 2024). Researchers and institutions from all over the world, including ecologists, plant biochemists, molecular biologists and physiologists, breeders, chemists, evolutionary biologists, and engineers, are expected to work together more actively to advance these next-generation technologies.

## 5.9 Conclusion

Climate change presents serious problems for food security, agriculture, and environmental sustainability, as well as for population persistence in natural communities. In order to assess the degree to which climate change disrupts eco-evolutionary processes and to forecast whether species can endure worsening conditions, an interdisciplinary approach utilizing ecological, evolutionary, and genomic tools is necessary to evaluate complex biological responses to climate change. Natural populations interact to climate change on an evolutionary level by changing their geographic ranges, adapting to new circumstances, or exhibiting a plastic reaction to shifting surroundings. Conserving and using biodiversity will be especially important for building food system resilience, if changing climates affect the stability of crop yields. These efforts depend on access to diverse forms of plant genetic resources.

## References

Akanwa AO, Mba HC, Jiburum U, Ogboi KC (2019) Strategies for combating climate change. In: Sustainable agriculture, forest and environmental management. Springer Singapore, Singapore, pp 393–435

Asbjornsen H, Hernandez-Santana V, Liebman M, Bayala J, Chen J, Helmers M, Ong CK, Schulte LA (2014) Targeting perennial vegetation in agricultural landscapes for enhancing ecosystem services. Renew Agric Food Syst 29(2):101–125

Asseng S, Ewert F, Martre P, Rötter RP, Lobell DB, Cammarano D, Kimball BA, Ottman MJ, Wall GW, White JW, Reynolds MP (2015) Rising temperatures reduce global wheat production. Nat Clim Chang 5(2):143–147

Audu V, Ruf T, Vogt-Kaute W, Emmerling C (2022) Changes in microbial biomass and activity support ecological intensification of marginal land through cultivation of perennial wheat in organic agriculture. Biol Agric Hortic 38(3):202–215

Brondízio ES, Aumeeruddy-Thomas Y, Bates P, Carino J, Fernández-Llamazares Á, Ferrari MF, Galvin K, Reyes-García V, McElwee P, Molnár Z, Samakov A (2021) Locally based, regionally manifested, and globally relevant: indigenous and local knowledge, values, and practices for nature. Annu Rev Environ Resour 46(1):481–509

Çekiç FÖ, Ünyayar S, Ortaş İ (2012) Effects of arbuscular mycorrhizal inoculation on biochemical parameters in Capsicum annuum grown under long term salt stress. Turk J Bot 36(1):63–72

Chadalavada K, Kumari BR, Kumar TS (2021) Sorghum mitigates climate variability and change on crop yield and quality. Planta 253(5):113

Chahal I, Hooker DC, Deen B, Janovicek K, Van Eerd LL (2021) Long-term effects of crop rotation, tillage, and fertilizer nitrogen on soil health indicators and crop productivity in a temperate climate. Soil Tillage Res 213:105121

Chen W, Wang J, Chen X, Meng Z, Xu R, Duoji D, Zhang J, He J, Wang Z, Chen J, Liu K (2022) Soil microbial network complexity predicts ecosystem function along elevation gradients on the Tibetan Plateau. Soil Biol Biochem 172:108766

Daryanto S, Wang L, Jacinthe PA (2016) Global synthesis of drought effects on maize and wheat production. PLoS One 11(5):e0156362

Das D, Sahoo J, Raza MB, Barman M, Das R (2022) Ongoing soil potassium depletion under intensive cropping in India and probable mitigation strategies. A review. Agron Sustain Dev 42(1):4

Deryng D, Conway D, Ramankutty N, Price J, Warren R (2014) Global crop yield response to extreme heat stress under multiple climate change futures. Environ Res Lett 9(3):034011

Diffenbaugh NS, Singh D, Mankin JS, Horton DE, Swain DL, Touma D, Charland A, Liu Y, Haugen M, Tsiang M, Rajaratnam B (2017) Quantifying the influence of global warming on unprecedented extreme climate events. Proc Natl Acad Sci 114(19):4881–4886

Dzienis G (2018) Winter rye-a species undervalued in Poland. A review. Agron Sci 73:19–28

Engler R, Randin CF, Vittoz P, Czáka T, Beniston M, Zimmermann NE, Guisan A (2009) Predicting future distributions of mountain plants under climate change: does dispersal capacity matter? Ecography 32(1):34–45

Fatichi S, Or D, Walko R, Vereecken H, Young MH, Ghezzehei TA, Hengl T, Kollet S, Agam N, Avissar R (2020) Soil structure is an important omission in Earth System Models. Nat Commun 11:522. [online]

Ferris H, Tuomisto H (2015) Unearthing the role of biological diversity in soil health. Soil Biol Biochem 85:101–109

Fitzgerald GJ, Tausz M, O'Leary G, Mollah MR, Tausz-Posch S, Seneweera S, Mock I, Löw M, Partington DL, McNeil D, Norton RM (2016) Elevated atmospheric [CO2] can dramatically increase wheat yields in semi-arid environments and buffer against heat waves. Glob Chang Biol 22(6):2269–2284

Ford JD, King N, Galappaththi EK, Pearce T, McDowell G, Harper SL (2020) The resilience of indigenous peoples to environmental change. One Earth 2(6):532–543

Garibaldi LA, Steffan-Dewenter I, Winfree R, Aizen MA, Bommarco R, Cunningham SA, Kremen C, Carvalheiro LG, Harder LD, Afik O, Bartomeus I (2013) Wild pollinators enhance fruit set of crops regardless of honey bee abundance. Science 339(6127):1608–1611

Gray SB, Dermody O, Klein SP, Locke AM, McGrath JM, Paul RE, Rosenthal DM, Ruiz-Vera UM, Siebers MH, Strellner R, Ainsworth EA (2016) Intensifying drought eliminates the expected benefits of elevated carbon dioxide for soybean. Nat Plants 2(9):1–8

Hawkins B, Sharrock S, Havens K (2008) Plants and climate change: which future? Botanic Gardens Conservation International, p 96 UK

Heinz M, Galetti V, Holzkämper A (2024) How to find alternative crops for climate-resilient regional food production. Agric Syst 213:103793

Hossain MS, Islam MN, Rahman MM, Mostofa MG, Khan MAR (2022) Sorghum: a prospective crop for climatic vulnerability, food and nutritional security. J Agric Food Res 8:100300

Jungers JM, DeHaan LH, Mulla DJ, Sheaffer CC, Wyse DL (2019) Reduced nitrate leaching in a perennial grain crop compared to maize in the Upper Midwest, USA. Agric Ecosyst Environ 272:63–73

Khatun M, Sarkar S, Era FM, Islam AM, Anwar MP, Fahad S, Datta R, Islam AA (2021) Drought stress in grain legumes: effects, tolerance mechanisms and management. Agronomy 11(12):2374

Kissoudis C, Van De Wiel C, Visser RG, Van Der Linden G (2016) Future-proof crops: challenges and strategies for climate resilience improvement. Curr Opin Plant Biol 30:47–56

Kleine T, Nägele T, Neuhaus HE, Schmitz-Linneweber C, Fernie AR, Geigenberger P, Grimm B, Kaufmann K, Klipp E, Meurer J, Möhlmann T (2021) Acclimation in plants–the Green Hub consortium. Plant J 106(1):23–40

Kopeć P (2024) Climate change—the rise of climate-resilient crops. Plants 13(4):490

Leisner CP (2020) Climate change impacts on food security-focus on perennial cropping systems and nutritional value. Plant Sci 293:110412

Loreau M, Naeem S, Inchausti P, Bengtsson J, Grime JP, Hector A, Hooper DU, Huston MA, Raffaelli D, Schmid B, Tilman D (2001) Biodiversity and ecosystem functioning: current knowledge and future challenges. Science 294(5543):804–808

Ma J, Cui B, Liu L, Hao X, Liang F, Jiang Z, Yang J (2022) Dynamic characteristics of drought conditions during the growth of winter wheat based on an improved SWAT model. Water 14(4):566

Mace GM, Norris K, Fitter AH (2012) Biodiversity and ecosystem services: a multilayered relationship. Trends Ecol Evol 27(1):19–26

Mondal R, Kumar A, Gnanesh BN (2023) Crop germplasm: current challenges, physiological-molecular perspective, and advance strategies towards development of climate-resilient crops. Heliyon 9(1)

Ottman MJ, Kimball BA, Pinter PJ, Wall GW, Vanderlip RL, Leavitt SW, LaMorte RL, Matthias AD, Brooks TJ (2001) Elevated CO2 increases sorghum biomass under drought conditions. New Phytol 150(2):261–273

Potts SG, Petanidou T, Roberts S, O'Toole C, Hulbert A, Willmer P (2006) Plant-pollinator biodiversity and pollination services in a complex Mediterranean landscape. Biol Conserv 129(4):519–529

Potts SG, Biesmeijer JC, Kremen C, Neumann P, Schweiger O, Kunin WE (2010) Global pollinator declines: trends, impacts and drivers. Trends Ecol Evol 25(6):345–353

Qiao L, Wang X, Smith P, Fan J, Lu Y, Emmett B, Li R, Dorling S, Chen H, Liu S, Benton TG (2022) Soil quality both increases crop production and improves resilience to climate change. Nat Clim Chang 12(6):574–580

Rebouh NY, Khugaev CV, Utkina AO, Isaev KV, Mohamed ES, Kucher DE (2023) Contribution of eco-friendly agricultural practices in improving and stabilizing wheat crop yield: a review. Agronomy 13(9):2400

Reilly EC, Gutknecht JL, Sheaffer CC, Jungers JM (2022) Reductions in soil water nitrate beneath a perennial grain crop compared to an annual crop rotation on sandy soil. Front Sustain Food Syst 6:996586

Reyes-Rosales A, Cabrales-Orona G, Martínez-Gallardo NA, Sánchez-Segura L, Padilla-Escamilla JP, Palmeros-Suárez PA, Délano-Frier JP (2023) Identification of genetic and biochemical mechanisms associated with heat shock and heat stress adaptation in grain amaranths. Front Plant Sci 14:1101375

Ricciardi V, Ramankutty N, Mehrabi Z, Jarvis L, Chookolingo B (2018) How much of the world's food do smallholders produce? Glob Food Sec 17:64–72

Rivero RM, Mittler R, Blumwald E, Zandalinas SI (2022) Developing climate-resilient crops: improving plant tolerance to stress combination. Plant J 109(2):373–389

Rogelj J, Meinshausen M, Knutti R (2012) Global warming under old and new scenarios using IPCC climate sensitivity range estimates. Nat Clim Chang 2(4):248–253

Rozema J, Aerts R, Cornelissen H (eds) (2007) Plants and climate change, vol 41. Springer Science & Business Media

Ryan MR, Crews TE, Culman SW, DeHaan LR, Hayes RC, Jungers JM, Bakker MG (2018) Managing for multifunctionality in perennial grain crops. Bioscience 68(4):294–304

Sanchez GM, Grone M, Apodaca A (2023) Indigenous stewardship of coastal resources in native California. Front Earth Sci 11:1064197

Satyavathi CT, Ambawat S, Khandelwal V, Srivastava RK (2021) Pearl millet: a climate-resilient nutricereal for mitigating hidden hunger and provide nutritional security. Front Plant Sci 12:659938

Sheehy J, Elmido A, Centeno G, Pablico P (2005) Searching for new plants for climate change. J Agric Meteorol 60(5):463–468

Sinha R, Zandalinas SI, Fichman Y, Sen S, Zeng S, Gómez-Cadenas A, Joshi T, Fritschi FB, Mittler R (2022) Differential regulation of flower transpiration during abiotic stress in annual plants. New Phytol 235(2):611–629

Tilman D, Isbell F, Cowles JM (2014) Biodiversity and ecosystem functioning. Annu Rev Ecol Evol Syst 45(1):471–493

Toor GS, Yang YY, Das S, Dorsey S, Felton G (2021) Soil health in agricultural ecosystems: current status and future perspectives. Adv Agron 168:157–201

Upadhyay SK, Srivastava AK, Rajput VD, Chauhan PK, Bhojiya AA, Jain D, Chaubey G, Dwivedi P, Sharma B, Minkina T (2022) Root exudates: mechanistic insight of plant growth promoting rhizobacteria for sustainable crop production. Front Microbiol 13:916488

Vejražka K, Kocourková Z, Faměra O, Sachambula L (2009) Grain physical properties of different rye varieties (Secale cereale) and baking quality Proceedings of the 5th International Congress Flour-Bread '09. 7th Croatian Congress of Cereal Technologists, Opatija, Croatia

Vijayan D, Ludwig D, Rybak C, Kaechele H, Hoffmann H, Schönfeldt HC, Mbwana HA, Rivero CV, Löhr K (2022) Indigenous knowledge in food system transformations. Commun Earth Environ 3(1):213

Waraich EA, Ahmed Z, Ahmad R, Ashraf MY, Naeem MS, Rengel Z (2013) 'Camelina sativa', a climate proof crop, has high nutritive value and multiple-uses: a review. Aust J Crop Sci 7(10):1551–1559

Wood A, Queiroz C, Deutsch L, González-Mon B, Jonell M, Pereira L, Sinare H, Svedin U, Wassénius E (2023) Reframing the local–global food systems debate through a resilience lens. Nat Food 4(1):22–29

Yunkaporta T (2019) Sand talk: how Indigenous thinking can save the world. Text Publishing

Zandalinas SI, Mittler R (2022) Plant responses to multifactorial stress combination. New Phytol 234(4):1161–1167

Zandalinas SI, Fritschi FB, Mittler R (2021) Global warming, climate change, and environmental pollution: recipe for a multifactorial stress combination disaster. Trends Plant Sci 26(6):588–599

# Chapter 6
# The Urban Gardening Movement

**Abstract** The future of global food production is full of obstacles. Global agriculture is under strain due to factors such growing urbanization, dwindling arable land, weather extremes brought on by environmental change, and a projected 9.6 billion people on Earth by 2050. Although more than 60% of the world's population already resides in cities, that percentage will increase to 70% by 2030. It's also important to consider global emissions. Currently, 20–30% of greenhouse gas emissions worldwide are caused by agriculture. Relocating food production to areas with strong demand lowers emissions and slows down global warming. Hence, the current book chapter highlights the importance of urban gardening and its impact on local food system and community health. It also provides practical advice on home gardening, container gardening, skyfarming and permaculture as a strategic pathway to mitigate hunger and food crisis in cities.

**Keywords** Container gardening · Food security · Home gardening · Permaculture · Roof gardening · Skyfarming

## 6.1 Introduction

The technique of growing crops and plants in urban areas is known as urban gardening. Although urban agriculture is often performed on a bigger scale, the terms are frequently used interchangeably. Food insecurity prompted the emergence of what are now known as "allotment gardens" in Germany throughout the 1800s. In order to alleviate the strain on food production and aid in war efforts, "liberty gardens" and "victory gardens" sprang up in the USA, Canada, Italy (known as "Orticelli di Guerra"), and the UK during World Wars I, II, and the Great Depression. Many were cultivated in the USA by immigrants from Europe, particularly Italy (Schmelzkopf 1995). Over 800 million people worldwide engage in urban agriculture today, despite the fact that many gardens, particularly in the Global North, are too small to provide enough food for the gardeners and their families on a daily basis. The media's craze for Michelle Obama cultivating a garden with school

 
N. Patel et al., *The Future of Food: Exploring Edible Plants for a Sustainable Diet*,
https://doi.org/10.1007/978-3-032-15027-1_6

children inside the White House is the example of how urban agriculture has gained significant political attention (Lyson 2004).

Most people are aware of the multiple benefits of gardening. First, by encouraging local, low-impact, and fresh food production for consumers in the area, urban agriculture helps to meet greenhouse gas emission targets. Even though gardening is sometimes done in highly contaminated soil and necessitates a lot of technical assistance, it improves the environmental quality of urban neighborhoods by reducing storm water runoff, filtering air and rainwater, reducing the effects of urban heat islands, acting as a sink for urban waste through decentralized composting, and assisting in the prevention of soil erosion. In addition to bringing greenery to previously devastated neighborhoods like Haddington in West Philadelphia, gardens are now growing throughout the city (Obach and Tobin 2014).

Thus, they frequently contribute to the beautification of locations. However, when increasingly appealing communities begin to regain their worth in the perspective of investors, community greening through urban cultivation carries hazards of gentrification and displacement. While the percentage of lower-income and gardeners of color has declined, urban gardens in cities like Delhi, New York, or Boston are in fact increasingly being maintained by newly arrived, wealthier citizens. Gardeners actively participate in garden production, preservation, and cleanup, which enhances and revitalizes local ties from a social perspective (McClintock 2010). They foster a stronger feeling of community and strengthen ties between residents and their neighborhood. In addition to sharing responsibilities and imagining a new use (beyond speculative usage) for the land, growers frequently construct a collaborative project without enclosing or claiming areas for private interests. Gardens encourage networking, social interactions, civic engagement, and a sense of community pride. From a health perspective, they provide people who might otherwise tend to isolate themselves at home alternatives for amusement and leisure as well as benefits for relaxation, healing, and trauma recovery (Hallett et al. 2016).

## 6.2 Why Cities Needs to Recognize Urban Farming as Hidden Gem?

Agriculture has long been thought of as existing in the rural realm, and in fact, its concomitant activities were typically limited to this environment. This has lead the way to the long-held belief that depending solely on the production of crops in rural areas would be adequate to feed the urban population. This proved to be somewhat incorrect for many cities in developing nations, primarily due to the lack of adequate infrastructure (roads, marketplaces, transportation etc.) and the low purchasing power of the underprivileged community (Drescher 2002).

In fact, the rise in poverty and high unemployment rates over time, combined with the opportunities that cities can provide—like the demand for food and the proximity to markets—have sparked the emergence of a number of cropping and food systems in cities and their environs, primarily focused on the production of fresh vegetables, milk, eggs and chickens (Ghosh 2002). The phrase "urban agriculture" was developed to refer to the raising of plants and animals in cities for domestic use and revenue. Additionally, the manufacturing and sale of agricultural inputs as well as the management and marketing of agricultural products after harvest are included in urban farming (Wekerle and Classens 2015).

Due of intense competition for land, it takes place in constrained areas. In urban agriculture, organic waste is recycled and city water is used. For this reason, it helps manage natural resources for a sustainable environment. The majority of farmers engaged in urban farming operations are characterized by a low degree of organization, and their harvest is sold fresh, without any additional processing. Urban agriculture currently supports rural production (mostly by supplying perishable goods like vegetables, milk, and eggs), and it is widely known that it enhances city food systems. For a number of cities, there is evidence of the growing importance of urban agriculture (Van Veenhuizen 2006). More than 21,000 hectares of Cagayan de Oro, Philippines, are utilized for urban agriculture; in Havana, Cuba, agriculture makes up over 12% of the urban land; and in Jakarta, Indonesia, more than 11,000 hectares are used for food production. Within the city of Harare (Zimbabwe), the final product seems to be highly varied and includes, the production of corn, vegetables, flowers and livestock (Cruz and Medina 2003).

Urban agriculture is expected to become an important component of most cities, both in advanced and developing nations, while many other activities are typically transitional. About 50% of Ghana's urban population is thought to work in agriculture in Accra, 80% in Brazzaville, Congo; 68% in Tanzania's five largest towns; 45% in Lusaka, Zambia; 37% in Maputo, Mozambique; 36% in Ouagadougou, Burkina Faso; and 35% in Yaoundé, Cameroon. These numbers unequivocally demonstrate how significant urban farming is to the informal sector's operations (Purnomohadi 2000). Although there are still some institutional resistance to the inclusion of urban agriculture in urban development plans, the number of initiatives (at the international, national, and local levels) addressing the promotion of urban agriculture has gradually expanded over the past 25 years. For example, in order to keep street criminals from hiding, agriculture was discouraged and for a long time outlawed in the peri-urban areas of Nairobi, Kenya, and many other African towns (Yi-Zhang and Zhangen 2000).

Conversely, governments in many nations encourage the growth of urban agricultural output. Since the early 1990s, new research endeavours have been conducted in this area as an outcome of growing awareness of the critical role that urban agriculture plays. For example, the International Development Research Centre (Canada) spearheaded global efforts to assist urban agriculture projects. Through the creative system-wide initiative known as "Urban Harvest," the Consultative group on International Agricultural Research (CGIAR) established urban agriculture as one of its research priorities in 2000 (Mougeot 2000). Since then, urban agriculture has been incorporated into the study programs of numerous National Study Centres in countries like Argentina, Kenya, Senegal, and Niger. Unexpected changes in urban life are being brought about by urban farming. It is about re-establishing human connections with nature and one another, not only about producing food. Growing food in urban areas can enhance wellbeing, rebuild and fortify the bonds between people and nature. Social groups and policymakers are taking advantage of this considering it as hidden gem (Prain et al. 2010).

## 6.3 Urban Farming and Its Impact on Local Food Systems and Community Health

Urban farming's primary strength is likely its contribution to food and nutrition security, as food production in cities gives those with lowest incomes easier access to food that meets a critical portion of their nutritional needs. The findings of a study that examined the relationship between urban agricultural involvement and nutritional adequacy using descriptive statistics and multivariate analysis verified that urban households can consume healthier, more nutrient-dense meals by farming (Maxwell et al. 1998; Zezza and Tasciotti 2010). Out of the 800 million people that work in urban areas globally, 200 million grow food for the market. Since the Urban Agriculture Network mostly relies on estimates derived from data extrapolation and individual experiences and observations, these figures are probably overstated. Actually, there is a dearth of quantitative research on urban agriculture that has been published, and a significant portion of the information provided is motivated more by advocacy goals than by rigorous analysis. The health of the poor in urban areas is improved by increased access to healthy food, specifically products that are fresh (Aubry et al. 2012). However, the safety of the public may be seriously threatened by what is grown in urban areas that are contaminated. Although this issue is still largely investigated, it shouldn't be disregarded, and as such, regulations and preventative measures ought to be strengthened. Every danger that is often present in conventional rural farming needs to be properly taken into account (Mutua et al. 2010).

The food supply system has recently been interrupted by the COVID-19 outbreak. Strengthening local food production at the household and community levels is therefore vital. The traditional method of transporting food across an average distance of 800–1500 km into major cities uses a lot of energy and is susceptible to disruptions like the one brought on by COVID-19 (Thiesen et al. 2022). Therefore, there is a pressing need to expand urban gardening in a methodical way in order to address some of the needs by adopting resilient food. Apart from the sizable and expanding megacities in developing nations, there are also a number of megacities in developed nations like Australia, Europe, and North America where this approach is just as essential for addressing the food crisis as it is in developing nations. In addition to lowering pressures on limited natural resources, scientific gardening and farming techniques have the ability to reduce food and nutritional insecurity and provide urban populations with a variety of sustainable, safe, and nourishing food options (McDougall et al. 2019).

The public's health is seriously threatened by pollution in cities that are growing quickly. Waste management is crucial and a major issue in the majority of cities across the world. By recycling inorganic wastes (for example, using plastic bottles, tanks, and automobile tires for soilless cultivation) as well as making compost from organic wastes, waste management in urban agriculture can improve the production of fresh food and maintain a clean urban environment (Despomier 2018).

## 6.4 Cultivating Urban Oases: Home Gardens, Containers and Permaculture in the Cities

### *6.4.1 Home Gardens*

The home garden is a minor scale production system that provides food for humans and animals as well as necessities that are either unavailable, prohibitively expensive, or easily obtained through field gardening, retail marketplaces, hunting, gathering, fishing, and wage labor. For reasons of convenience, security, and extra attention, household gardens are typically situated close to homes. They occupy labor that is peripheral to important household economic activities and land that is peripheral to field output (Odebode 2006). Household gardens with ecologically adapted and complimentary plants are distinguished by their modest capital input and straightforward technology. These gardens are a common feature of small-scale subsistence farming systems in both rural and urban regions. The origins of contemporary agriculture can be traced back to subsistence farming practices, which started in minute garden plots scattered around homes (Kumar and Nair 2004). These gardens have steadfastly withstood the time and are crucial to the family's consumption of food and revenue. Since the Dutch researchers Ochse and Terra (1934) conducted the first home garden studies in the 1930s on mixed gardens in Java, Indonesia, there have been multiple contributions in the field that synthesize definitions, species inventories, functions, structural features, composition, socioeconomic significance, and cultural significance.

In addition to microgreens, fruits, spices, herbs, ornamental plants, medicinal plants and livestock that can be used as a supplemental source of food and income, home gardens can also be defined as a mixed cropping system. Five inherent qualities of home gardens were enumerated by Michell and Hanstad (2004): (1) Must be situated close to homes; (2) Must have wide variety of plants; (3) The production serves as a supplement to family earnings and consumption rather than as their primary source; (4) Require almost little space; and (5) Constitute a system of production that the underprivileged may readily participate in to some extent.

The two biological types of these gardens are temperate and tropical. Home gardens in Central and South American tropical regions are the subject of a large portion of the literature. Home gardens are also very popular in South and Southeast Asia as well as Africa. Consequently, there aren't many published studies on home gardens from countries with large populations and temperate climates (Méndez et al. 2001). A home garden's exact dimensions vary from one household to the next, although often, they are less than the household's arable land. Nevertheless, this might not be the case for landless people and families without agricultural land. Home gardening is now feasible even for families with very little or no land owing to new developments and technologies. Physical demarcations, such as living hedges, ditches, fences or boundaries developed through mutual consent, can be used to define the limits of home gardens. Home gardeners have long applied animal manure, kitchen scraps and other organic residues, and this technique has significantly improved the fertility and production of these gardens (Aguilar-Støen et al. 2009).

Growing a home garden may be extremely enjoyable. The needs of the household in terms of consumption and earning income largely influence decisions about crop selection, input procurement, harvesting, management and other concerns. According to an Indonesian study, the socioeconomic standing of the household can have an impact on the composition, structure and diversity of home gardens (Abdoellah et al. 2006). For example, when the families established themselves financially, they started growing horticulture crops instead of staples, and some of them started rearing livestock. Even though home gardening requires less horticultural and agronomic expertise, crop failures and other adverse effects can be minimized when family members are equipped with greater knowledge and abilities (Galhena et al. 2013).

### *6.4.2 Container Gardens*

The practice of growing all plants, including herbs and vegetables, in containers is known as container gardening. Despite the quick advancements in cities, people find it difficult to come up with options for growing food and to supply some alternatives to increase hunger during a pandemic. To prevent the worst consequences for nutrition and food security, the community must continue to closely monitor the situation and take appropriate action and should look at factors that could affect food security, such as use, stability, accessibility, and use (Deveza and Holmer 2002). Those who resided in urban areas and faced food insecurity included those who had been displaced by armed conflicts or natural disasters. Food insecurity is becoming a global concern. In order to maintain food security throughout the outbreak of the disease, farming in cities expanded faster. Even after the pandemic, meeting the demands of the steadily growing urban population and reducing the burden on rural agriculture became imperative. People just don't need food; it's all about ingesting the appropriate kind of nutrient-dense food. It was discovered that one in three individuals, primarily youngsters, were underweight. Early deprivation of vital nutrients can cause malnutrition and other problems in children. People never understood the need of food as long as they could eat on an empty stomach. Healthy food must be available, uninterrupted, and consistent at all times (Kaur et al. 2024). According to earlier research, when someone encounters a difficult life circumstance, they experience stress, which subsequently has a detrimental impact on their health and general well-being.

With the world's fastest-aging population, Japan in particular needs to better control its growing social security and national health care expenses. In addition to enhancing physical health through exercise, consideration should be given to the emotional and social components of well-being. It has been said that, in light of these kinds of societal demands, city design can significantly contribute to the improvement of public health (Nagase and Lundholm 2021). One initiative that has received an enormous amount of attention is providing urban farmers and gardeners with the opportunity to engage in these activities. Additionally, container gardening combats global climate change and domestic hunger. It also teaches people practical skills and reconnects them with nature. If these methods are adopted and perpetuated, the Filipino working class may never again face hunger in the same manner as much as possible by those who require a route from hunger to wealth (Denzon et al. 2023).

The crucial elements of container gardening includes selecting containers: The containers that are suitable for the plants you wish to cultivate must be selected by focusing on the elements like dimensions, composition (such as wood, plastic or clay) drainage holes, and visual attractiveness. To avoid waterlogging, make sure containers have enough drainage. Additionally, take durability, insulation, and root space requirements into account. Choose plants that are appropriate for container gardening. Take into account their size, growth pattern, and need for sunlight. Container gardening is a popular way to grow flowers, vegetables, herbs, miniature shrubs and even dwarf fruit trees. Because of their lower size and greater container adaptation, dwarf or compact cultivars are frequently preferred (Nwosisi and Nandwani 2018). A premium potting soil or a growing medium that drains properly and is made especially for container gardening can be used. Garden soil should not be used because it can be too compacted and heavy in containers. The soil should hold onto moisture while efficiently letting extra water runoff. On the other hand, nutrients may drain out more quickly from container plants, they may need to be fertilized on a frequent basis. Depending on the particular requirements of the plant, application of liquid fertilizer or a slow-release fertilizer can be implemented. It is essential to regularly check the pH and fertility of the soil for better plant growth (Paraskevopoulou 2021).

For optimal production, the containers can be placed in areas that receive the right amount of daylight. While some plants prefer partial shade, others need full exposure to the sun. The elements like the shadows of surrounding trees and buildings could be averted in order to maintain optimum growth (Feldmann and Vogler 2021). Management of pest and disease is also important and for that the use of integrated pest management strategies, such as organic pest treatment, companion planting, and routine monitoring can be followed. Planting flowers along with herbs will draw pollinators and beneficial insects, promoting biodiversity. There are several benefits to container gardening, such as the capacity to regulate growing conditions and provide mobility and adaptability. It is a popular option for city people and those without access to traditional gardens because it enables gardening aficionados to enjoy plants and grow their own food even in small places (Shepherd 2011).

### 6.4.3 Permaculture

The growing realization that the environment was in crisis during the 1960s and 1970s gave rise to permaculture. During that period, ecology was maturing as a distinct field. In order to establish "a design system for creating sustainable human environments," Bill Mollison and David Holmgren developed "permaculture," an applied ecology system that stemmed from "observation of ecological systems, the wisdom embedded in traditional farming systems, and modern technological as well as scientific knowledge." In order to manage land and plan and build homes, gardens, farms, and communities in a sustainable and equitable manner, permaculture encompasses twelve design principles and three ethics. Permaculture principles include sharing excess resources, taking care of people, and protecting the environment. Permaculture differs from both the ethnocentric, cultural ecology perspective and the ecocentric, fundamental ecology perspective in that it places equal focus on environmental preservation and human well-being.

Holmgren (2002) revisited the concepts, theories, and practices of permaculture and reformulated the principles suggested by Mollison et al. (1991) which emphasized design elements and their complementary interaction within landscapes, reducing energy use and utilizing biological resources (Fig. 6.1). Making significant adjustments to many aspects of society and decision-making will be needed to move cities toward sustainability. An integrated strategy for implementing these changes and involving all societal levels has been offered by permaculture. The planning and regulatory bodies of the local and state governments, the development sector, and the planning and design experts that work for these organizations are important stakeholders who determine and impact the paths that urban areas take as they grow and change (Taylor Aiken 2017). Permaculture can be broadly categorized as a field of ecological planning and environmental engineering that seeks to create self-maintained agricultural systems and sustainable human settlements that are modeled after natural ecosystems. A Japanese philosopher and farmer named Masanobu Fukuota, developed an ecological farming method known as "natural farming," which some have claimed includes the fundamental elements of permaculture. There are other names for the technique, such as "the Fukuoka Method," "the natural way of farming," or "do-nothing farming," which refers to minimizing manufactured inputs and equipment rather than labor shortages (Akhtar et al. 2016).

The system takes advantage of how each ecosystem is shaped by the intricacy of living things. Fukuoka regarded farming as an artistic or spiritual way of living, with "the cultivation and perfection of human beings" as its ultimate purpose, rather than just a way to produce food. In contrast to traditional organic farming, which Fukuoka believed to be another contemporary method that disrupts nature, natural farming is a closed system that mimics nature and requires no inputs. According to Fukuoka, his method avoids soil erosion, biodiversity loss, and water pollution without sacrificing the quantity of food it can produce (De Tombeur et al. 2018). Similar assertions have been advanced by contemporary permaculturists, who even assert that the technique can provide higher yields per hectare than traditional farming. However, we must keep in mind that contemporary farming has only been

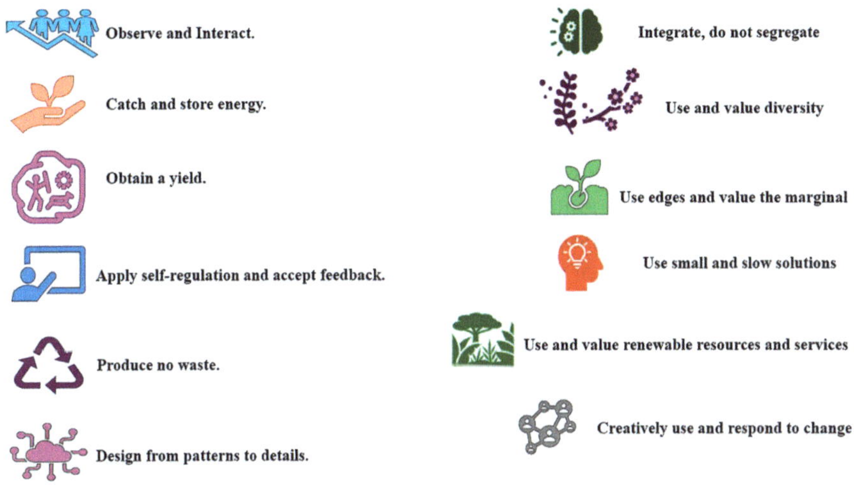


**Fig. 6.1** Principles of Permaculture

"conventional" for roughly 60 years! One significant change brought about by the shift to permaculture is that we must modify our diet to include more fruits, vegetables, nuts, berries, and other foods that can be produced efficiently using its methods because grains cannot be produced at the scale of industrialized agriculture (Fiebrig et al. 2020).

The aim of permaculture is to maintain an efficient design that prioritizes species assembly, function, and landscape patterns. It poses the following queries: Where does this component go? How may it be arranged in relation to other components to maximize the system's overall benefit? Permaculture's core idea is to maximize beneficial linkages between elements in order to maximize their synergy in the finished, ideal design. Instead of emphasizing individual components in isolation, permaculture emphasizes the connections made between them through their placement; the entirety is superior than the average of its parts. In order to attain a high degree of holistic integrity and resilience, permaculture design seeks to minimize waste, human labor, and inputs of energy and other resources through the development of systems with maximum benefits between design parts. Permaculture designs are "natural" and change over time with respect to the interactions between these components and linkages. They can develop into incredibly complex systems that can generate a large amount of food and materials with little input (Didarali and Gambiza 2019).

### *6.4.4 Skyfarming*

Skyfarming is a cutting-edge method of producing crops indoors in specially designed, multi-story structures to meet the rising need for basic foods while halting the widespread conversion of natural ecosystems into agricultural land. The concept seeks to provide a highly productive and resource-efficient food production system with ideal growing conditions and no seasonal changes within a technically optimized building envelope. Currently, 159 million hectares of rice are being grown, accounting for more than 22% of the world's grain producing area. At the same time, rice irrigation accounts for 24–30% of total human water use. The total ratio of water used for rice cultivation per year between 1998 and 2001 was 941 $km^3$. It was predicted that 15–20 million hectares of rice would experience both financial and physical water limitations starting in 2027 as a result of heightened competition for water and the growth of the farmed area. In order to feed the world's expanding population, rice yields must steadily rise over the coming decades despite the depletion of land and water resources (Putri et al. 2015). Skyfarming employs a cost-effective approach, which minimizes the amount of resources needed to produce one unit of rice. This efficiency boost will be accomplished by technological advancements that maximize growing conditions, enhance the amount of biomass used, and provide a high level of recycling of energy, water, and nutrients. It is intended to relocate the plants to regions where the ideal conditions for each growth stage are most effectively developed and preserved with the least amount of energy input in order to lower the demand for both energy and supplies (Datta et al. 2017). In place of hydroponics or hefty, challenging-to-handle rooting material, nutrient-rich mist will give the roots the necessary water and nitrogen supply. The amount of water required to be moved throughout the structure will be greatly decreased by this innovative technology, which will also lessen its statistical needs. Only a small portion of the land, water, and fertilizers needed for traditional agriculture will be needed for skyfarming. Additionally, it will be mostly resistant to the consequences of climate change (Siebert and Döll 2010).

### *6.4.5 Green Roof Farming*

These days, urban heat islands and a shortage of green space are issues in numerous cities across the world. By enhancing urban vegetation and encouraging urban gardening or agriculture, some cities are attempting to improve sustainability. Green roofs combined with farming in cities can help mitigate environmental issues, improve community functions, and develop urban food systems, making it feasible to achieve environmental, social, and economic sustainability for buildings in urban areas (Aboulnaga and Fouad 2021). Installing plants that grow on roofs is known as a "green roof system." Through the provision of green spaces, mitigation of temperature, reduction of air quality issues, improvement of storm water management, and biodiversity, they could favorably contribute to making cities more livable. In reality, green roofs offer a variety of advantages, including financial, technical, ecological, and recreational benefits. Green roofs are thought to improve local human climatic conditions as well as cities' aesthetic appeal (Hashim et al. 2020).

## 6.5 Navigating Challenges for Sustainable Success

Competition from non-agricultural economic activities, habitat, transportation, etc., is fierce in urban farming.In addition to labor, there is fierce rivalry for land and inputs like fertilizer and water. Additionally, transportation, domestic activities like heating and cooling the home and workplace, and industry all contribute to the high levels of pollution found in urban and peri-urban areas. However, it is well recognized that the use of chemical and organic fertilizers, as well as pesticides, pollutes the environment. Therefore, the challenge for urban farming is to show that, in spite of an occasionally polluted urban environment, it generates safe food items rather than polluting the city environment (de Bon 2003). Despite numerous attempts to boost yield, establish pest and disease-resistant cultivars, and develop methods for small-scale farming, the main inputs utilized in agricultural production are fertilizer and water. Nearby cities may offer chances to obtain a portion of them through the utilization of solid and liquid waste products, the availability of which rises as cities expand (Meynard et al. 2001).

Solid municipal waste, vegetable compost, animal waste, and chemical fertilizers are among the numerous sources of nutrients used in urban crop production. Every urban farmer in every city uses chemical fertilizers. Fish ponds are another application for solid waste. Despite being used frequently; organic matter is not used as widely. Organic waste is either composted or used fresh. In the cities of emerging nations, solid waste is frequently used for urban agriculture. Paper and kitchen waste make up the majority of the garbage in these cities, making up 42% and 19% of Metro Manila's total waste, respectively. These waste materials have a relatively low nutrient content; for instance, the organic garbage in Ougadougou and Bamako only contains 0.29% nitrogen and 0.16% phosphorus (Anh et al. 2004). Many initiatives have been put into place to promote the use of various waste materials from municipal projects by setting up compost plants for individual and community growers in urban areas employing trench composting, vermiculture, compost chambers, containers and heaps. The health concerns to consumers and farmers handling the compost must be considered while composting municipal solid waste (Akinbamijo et al. 2002). This covers zoonosis, disease vectors, the persistence of harmful organisms (*Salmonella, E. coli,* and *B. cereus*), and chemical pollution from heavy metals and persistent organic chemicals. To reduce these hazards, trash must be sorted using a house-to-house source-separated garbage collection system with an effective composting process for the right raw materials. Cities in emerging nations like China and Mexico, as well as those in underdeveloped nations, use wastewater for fisheries and cultivation of crops including ornamentals, vegetables, tree fruits, and feed. Thus, the utilization of solid and liquid wastes presents a chance to improve crop production and purify the contaminated urban environment. Although there are still issues with human health, these wastes may provide all or some of the nutrients required for urban gardening (Cofie et al. 2006).

The metropolitan surrounding air, water supplies, and soils have all become contaminated. According to analyses, soils in urban areas are more contaminated than those in rural ones. In one study, all urban soil samples had organic pollutant (benzo(a)pyrene) concentrations more than 0.06 $mg.kg^{-1}$, while only 15% of the soil samples came from rural areas. Numerous investigations on polychlorobiphenyls and polycyclic aromatic hydrocarbons pollutions have found similar results. Heavy industries and runoff via roadway drains are the main causes of this contamination (Lungu 2002). In Asian cities, this kind of pollution is also frequently discovered in irrigation and aquaculture water. Transportation, home heating, and industry are the main causes of air pollution. In metropolitan areas, $SO_2$ and $CO_2$ concentrations are greater than the permissible range. The use of artificial chemical pesticides is a significant barrier to the growth of agriculture in and around cities (Cissé et al. 2002).

Regular pesticide applications are generally advised by technical advisors for the production of vegetable, ornamental, and flower crops. Urban vegetable production frequently uses a variety of active compounds from all the major chemical families, including carbamates, pyrethroids, organochlorines, and organophosphates. There is still more work to be done in educating farmers, extension agents, and chemical retailers and traders about identifying pests and diseases, applying pesticides correctly, and promoting less toxic pesticides, even with the many initiatives on Integrated Production Management in large-city urban agriculture around the world (De Bon et al. 2010).

## 6.6 Conclusion

The demand for products generated from plants continues to increase due to the world's expanding population. Land availability and production intensity are the two main constraints on conventional crop production. Exhaustion of production resources, environmental degradation, and biodiversity loss are just a few of the serious and frequently irreversible effects that could result from pushing land expansion and/or intensification to the limit. Thus, Urban gardening plays an important role in providing scientific and technical knowledge for sustainable diet and environment.

## References

Abdoellah OS, Hadikusumah HY, Takeuchi K, Okubo S, Parikesit (2006) Commercialization of homegardens in an Indonesian village: vegetation composition and functional changes. Agrofor Syst 68(1):1–13

Aboulnaga M, Fouad H (2021) Urban green coverage: importance of green roofs and urban farming policies in enhancing liveability in buildings and cities—global and regional outlook. In: The importance of greenery in sustainable buildings. Springer International Publishing, Cham, pp 155–230

Aguilar-Støen M, Moe SR, Camargo-Ricalde SL (2009) Home gardens sustain crop diversity and improve farm resilience in Candelaria Loxicha, Oaxaca, Mexico. Hum Ecol 37(1):55–77
Akhtar F, Lodhi SA, Khan SS, Sarwar F (2016) Incorporating permaculture and strategic management for sustainable ecological resource management. J Environ Manag 179:31–37
Akinbamijo OO, Fall ST, Smith OB (2002) The production environment of the horticulture-livestock integration. Option in Sénégambia urban agriculture. In: Advances in crop-livestock integration in West African cities, ITC, ISRA, CRDI, Ottawa, Canada, pp 37–52
Anh MTP, Ali M, Anh HL, Ha TTT (2004) Urban and peri-urban agriculture in Hanoi: opportunities and constraints for safe and sustainable food production. AVRDC-World Vegetable Center
Aubry C, Ramamonjisoa J, Dabat MH, Rakotoarisoa J, Rakotondraibe J, Rabeharisoa L (2012) Urban agriculture and land use in cities: an approach with the multi-functionality and sustainability concepts in the case of Antananarivo (Madagascar). Land Use Policy 29(2):429–439
Cissé I, Fall ST, Akinbamijo OO, Diop YM, Adediran SA (2002) L'utilisation des pesticides et leurs incidences sur la contamination des nappes phréatiques dans la zone des Niayes au Sénégal. In: Advances in crop-livestock integration in West African cities, pp 85–100
Cofie O, Bradford A, Drechsel P (2006) Recycling of urban organic waste for urban agriculture. In: Cities farming for the future: urban agriculture for green and productive cities, pp 207–229
Cruz MC, Medina RS (2003) Agriculture in the city: a key to sustainability in Havana, Cuba. Ian Randle Publishers
Datta A, Ullah H, Ferdous Z (2017) Water management in rice. In: Rice production worldwide. Springer International Publishing, Cham, pp 255–277
de Bon H (2003) Improving techniques for peri-urban vegetable production in South-East Asia. Urban Agr- Mag 10:12–13
De Bon H, Parrot L, Moustier P (2010) Sustainable urban agriculture in developing countries. A review. Agron Sustain Dev 30(1):21–32
De Tombeur F, Sohy V, Chenu C, Colinet G, Cornelis JT (2018) Effects of permaculture practices on soil physicochemical properties and organic matter distribution in aggregates: a case study of the bec-hellouin farm (France). Front Environ Sci 6:116
Denzon W, Estose MC, Villanueva KC, Jao JH (2023) Container gardening program as a way of food augmentation: action research. Multidiscip Sci J 5(4):2023043–2023043
Despomier D (2018) Vertical farming using hydroponics and aquaponics. In: Urban Soils, pp 313–327
Deveza KS, Holmer RJ (2002) Container gardening: a way of growing vegetables in the city. Food and Agriculture Organization, Rome
Didarali Z, Gambiza J (2019) Permaculture: challenges and benefits in improving rural livelihoods in South Africa and Zimbabwe. Sustainability 11(8):2219
Drescher AW (2002 September) Food for the cities: urban agriculture in developing countries. International conference on urban horticulture 643:227–231
Feldmann F, Vogler U (2021) Towards sustainable performance of urban horticulture: ten challenging fields of action for modern integrated pest management in cities. J Plant Dis Prot 128(1):55–66
Fiebrig I, Zikeli S, Bach S, Gruber S (2020) Perspectives on permaculture for commercial farming: aspirations and realities. Org Agric 10(3):379–394
Galhena DH, Freed R, Maredia KM (2013) Home gardens: a promising approach to enhance household food security and wellbeing. Agric Food Secur 2(1):8
Ghosh S (2002) Food production in cities. international conference on urban horticulture 643:233–239
Hallett S, Hoagland L, Toner E (2016) Urban agriculture: environmental, economic, and social perspectives. Hortic Rev 44:65–120
Hashim NH, Mohd Hussain NH, Ismail A (2020) Green roof concept analysis: a comparative study of urban farming practice in cities. Malays J Sustain Environ (MySE) 7(1):115–132
Holmgren D (2002) Permaculture: principles and pathways beyond sustainability. Permanent Publications, Hampshire. [online]

Kaur A, Ratan P, Sharma A (2024) Urban kitchen gardening: realization of its importance during the COVID-19 pandemic. In: Food Security and Sustainability, pp 275–285
Kumar BM, Nair PR (2004) The enigma of tropical homegardens. Agrofor Syst 61(1):135–152
Lungu M (2002 June) Issues of urban soil and water management in relation to urban agriculture in CEE/NIS countries. In: Urban agriculture and cities in transition. Proceedings of the regional workshop, pp 20–22
Lyson TA (2004) Civic agriculture: reconnecting farm, food, and community (Civil Society). Tufts University
Maxwell D, Levin C, Csete J (1998) Does urban agriculture help prevent malnutrition? Evidence from Kampala. Food Policy 23(5):411–424
McClintock N (2010) Why farm the city? Theorizing urban agriculture through a lens of metabolic rift. Camb J Reg Econ Soc 3(2):191–207
McDougall R, Kristiansen P, Rader R (2019) Small-scale urban agriculture results in high yields but requires judicious management of inputs to achieve sustainability. Proc Natl Acad Sci 116(1):129–134
Méndez VE, Lok R, Somarriba E (2001) Interdisciplinary analysis of homegardens in Nicaragua: micro-zonation, plant use and socioeconomic importance. Agroforest Syst 51(2):85–96
Meynard JM, Doré T, Habib R (2001) The evaluation and design of cropping systems for sustainable agriculture. C R Acad Agric Fr 87(4):223–236
Mitchell R, Hanstad T (2004) Small homegarden plots and sustainable livelihoods for the poor. FAO LSP WP, p 11
Mollison BC, Slay RM, Girard JL, Girard JL (1991) Introduction to permaculture [online]
Mougeot LJ (2000) Urban agriculture: definition, presence, potentials and risks, and policy challenges. Cities feeding people series; rept. 31 International Development Research Centre (IDRC) Ottawa, Canada
Mutua GK, Kinyari P, Githuku C, Kironchi G, Kang'ethe E, Prain G, Njenga M, Karanja NN (2010) Assessment of environmental and public health hazards in wastewater used for urban agriculture in Nairobi, Kenya. Trop Subtrop Agroecosystems 12:85–97
Nagase A, Lundholm J (2021) Container gardens: possibilities and challenges for environmental and social benefits in cities. J Living Architect 8(2):1–19
Nwosisi S, Nandwani D (2018) Urban horticulture: overview of recent developments. In: Urban horticulture: sustainability for the future. Springer, Cham, pp 3–29. https://doi.org/10.1007/978-3-319-67017-1_1
Obach BK, Tobin K (2014) Civic agriculture and community engagement. Agric Hum Values 31(2):307–322
Ochse JJ, Terra GJA (1934) Research on the economic and agricultural condition and food consumption in Koetowinangoen. Landbouw. The Netherlands
Odebode SO (2006) Assessment of home gardening as a potential source of household income in Adinyele Local Government are of Oyo State. Niger J Hortic Sci 11(1):47–55
Paraskevopoulou A (2021) Horticulture, design, and ecology: how to deal with the urban environment? In: VIII international conference on landscape and urban horticulture, vol 1345, pp 1–12
Prain G, Karanja N, See-Smith D (2010) African urban harvest: agriculture in the cities of Cameroon, Kenya and Uganda. IDRC/Springer, Ottawa, ON, CA/Lima
Purnomohadi N (2000) Jakarta: urban agriculture as an alternative strategy to face the economic crisis. In: Bakker N, Dubbeling M, Guendel S, Sabel Koschella U, de Zeeuw H (eds) Growing cities, growing food, urban agriculture on the policy agenda. DSE, Feldafing, pp 453–466
Putri NY, Sharfina NP, Prakarti T (2015) Sky farming: the alternative concept of green building using vertical landscape model in urban area as an effort to achieve sustainable development. Int J Civil Environ Struct Construct Architect Eng 9(7):938–941
Schmelzkopf K (1995) Urban community gardens as contested space. Geogr Rev 85:364–381
Shepherd L (2011) The complete guide to growing vegetables, flowers, fruits, and herbs from containers: everything you need to know explained simply. Atlantic Publishing Company

Siebert S, Döll P (2010) Quantifying blue and green virtual water contents in global crop production as well as potential production losses without irrigation. J Hydrol 384(3–4):198–217

Taylor Aiken G (2017) Permaculture and the social design of nature. Geogr Ann Ser B, Hum Geogr 99(2):172–191

Thiesen T, Bhat MG, Liu H, Rovira R (2022) An ecosystem service approach to assessing agroecosystems in urban landscapes. Land 11(4):469

Van Veenhuizen R (2006) Cities farming for the future. In: Cities farming for future, urban agriculture for green and productive cities. RUAF Foundation, IDRC and IIRP, ETC-Urban agriculture, Leusden, pp 2–17

Wekerle GR, Classens M (2015) Food production in the city:(re) negotiating land, food and property. Local Environ 20(10):1175–1193

Yi-Zhang C, Zhangen Z (2000) Shanghai: trends towards specialised and capital intensive urban agriculture. In: Growing cities growing food: urban agriculture on the policy agenda. DSE-ZEL, Feldafing, pp 467–477

Zezza A, Tasciotti L (2010) Urban agriculture, poverty, and food security: empirical evidence from a sample of developing countries. Food Policy 35(4):265–273

# Chapter 7
# Cultural Perspectives on Edible Plants

**Abstract** Three-quarters of the world's population and public health interests are found in developing nations, which most closely resemble the legacy of our traditional forefathers. In the developing world, the majority of diets are plant-based for cultural, religious, traditional, and economic reasons. High rates of vitamin A, iron, zinc, riboflavin, and vitamin B12 deficient states are linked to this. Chronic energy shortage and poor linear growth occur simultaneously. On the other hand, except for cataract, osteoarthritis, and gastro-esophageal cancer, the major chronic diseases are not very common among tribal groups and rural peasants. A plant-based diet regimen offers the largest content of inhibitors and the lowest content of promoters of metabolic dysregulation, which are the main causes of disease and debility in adults over 40. This is true regardless of the "original" dietary pattern of pre-agricultural humanity. Maintaining and strengthening traditional eating habits while enhancing micronutrient delivery and avoiding any negative environmental effects from traditional cooking methods is a challenge for developing nations. Hence, the present chapter highlights the types of cuisine and cultural essence of diverse traditions and countries along with their contributions to health and sustainability.

**Keywords** Cultural essence · Cuisine · Diversity of food · Globalization · Health and sustainability

## 7.1 Introduction

Cultural cuisine is defined as the repository of food-related customs that have been developed and passed down through an ethnic group. Together with the symbolic food combinations in meals and menus, as well as seasonal or lifetime traditions of ritual foods and eating, it covers the selected basic foods, distinctive flavorings, processing (such as chopping and cooking), regulations pertaining to permissible foods, combinations, and textures, festival foods, and the social context of eating. The idea encompasses diet, which includes the kinds and amounts of food and drink as well as their role in macro and micronutrient intakes, and gastronomy, which is



N. Patel et al., *The Future of Food: Exploring Edible Plants for a Sustainable Diet*,
https://doi.org/10.1007/978-3-032-15027-1_7

the art and science of healthy eating (Messer 1984). The foods that people consume and the ways that they cook and consume them vary greatly throughout the world. How did certain cuisines get at their current status? The main factors influencing staple foods include soil, climate, and historical plant and animal domestication and subsequent global dispersion. However, cuisine might not be influenced by the foods that are nearby. For instance, there would be significant variations in the food that chefs from China, France, and Colombia developed if they were allowed to freely select from the enormous food markets in Hangzhou or other region (Albala 2000).

The methods of preparation, cooking, and serving would differ greatly even if they selected the same fundamental set of ingredients. Chinese restaurants all across the world demonstrate how well Chinese chefs can adjust their recipes to suit local ingredients. Although their food frequently reflects regional tastes, it is nevertheless clearly Chinese. Similar to other facets of culture like language, regional differences in food exist among nations. As development progresses, regional variations in cuisine also tend to diminish (Wahlqvist et al. 1997).

Farmers have largely supplanted the hunter-gatherers of 15,000 years ago, with the exception of regions with extremely little rainfall, which are home to nomadic cattle herders. The ability to settle in villages and eventually unite into bigger groups was made possible by the domestication of wildlife and plants. Through trade, plants were introduced to comparable climates throughout Asia. Many new foods were brought to Eurasia and Africa by European explorers during their exploration of the Americas. Following Columbus, the "Mediterranean Diet" underwent significant change (Milton 2000). The concept of southern Italian food without tomatoes, for instance, which was introduced from the Americas, now seems remarkable. The complexity of social organization within regions has influenced the evolution of cuisines; but, in regions where communication barriers, whether physical or the result of conflict, existed, cuisines emerged separately. As a result of globalization, agribusiness, and urbanization, cuisine keeps changing (Renna et al. 2015).

Animal domestication was based on what was appropriate for a given location and what was accessible. The Rendille herd primarily consists of camels, but the Samburu have mostly cattle in the semi-desert region of northern Kenya. Since then, a mission that provides a steady supply of water from wells has upended the Rendille's traditional adaption to their surroundings. Subsequently, overgrazing led to significant environmental damage and a reliance on non-traditional crops (Smith 2005). In the nineteenth and twentieth centuries, many colonized populations experienced widespread disruption of ecological systems and human adaptations, leading to physical and social sickness. This phenomenon persists today in every continent worldwide, particularly among former hunter-gatherers and animal herders whose diets have undergone significant change. For instance, the loss of their traditional cuisine has led to a high death rate from chronic diseases among Australia's erstwhile hunters and gatherers aborigines (Smith and Smith 2003). The present chapter highlights the types of cuisine and cultural essence of diverse traditions and countries along with their contributions to health and sustainability.

## 7.2 Global Culinary Diversity: Exploring the Types of Cuisine

There are differences in eating patterns between societies as well as occasionally within a single community. A food type that is forbidden in one culture could be considered normal in another. Furthermore, some people become interested in these forbidden foods and travel to such areas to sample them. The diversity of civilizations and their differences have influenced the development of the present tourist profile.

The cuisines are mostly characterized on the basis of historical, geographical, diet types and origin. The historical component that is used to identify the different varieties of cuisine pertains to the emergence of culinary and food consumption customs and practices from the inception of cooking to the present day. The growth of food preparation, cooking, and serving methods has a strong connection to the historical component. The style is therefore one of the most essential aspects when this issue is being investigated. Additionally, historical events have always been associated with certain areas and regions, the geographical aspect is another crucial consideration. This feature can be applied on a variety of scales, ranging from a single nation to an entire continent. The Hellenistic (ancient Greek) cuisine served as the model for ancient cuisine, and the Roman cuisine emerged afterwards (Anderson 2014). Europe's culinary traditions are thought to have originated with Hellenistic food. The "Mediterranean triad"—wheat, olive oil, and vine—forms the foundation of this cuisine. The Hellenistic culinary philosophy became a fundamental principle of Roman cuisine. The division of classes is the one feature that all of the European nations share. Large and lavish fasts, where the focus was on an abundance of food in terms of both type and quantity, were one method for wealthy individuals to distinguish themselves from those who were struggling. Classical cuisine is the term used to describe the cuisine that emerged at the end of the nineteenth century and the start of subsequent ones. This cuisine is still consumed today, and imitations of it are widely accepted and dispersed globally. This cuisine was referred to as haute (high) cuisine during its inception (Counihan and Van Esterik 2013).

Cuisine in an area in particular is defined by geographical factors. A nation, a region, or an entire continent may be included in this territory. The national cuisine of a nation is discussed when discussing its area. This factor takes into account that a national cuisine might be further subdivided according to the country's geographical sub regions. The regional component encompasses numerous national cuisines of different cultures whose culinary customs and traditions share many traits (Wahlqvist 2007). These typical traits are frequently dictated by the environment, which affects the items that can be produced in a certain area (fruits, vegetables, livestock, and so forth). Mediterranean cuisine, which encompasses the culinary traditions of Greece, Italy, and Spain—and oriental cuisine serve as excellent illustrations of this. As the term implies, when a cuisine is compared across continents, it incorporates the most prevalent elements of that continent's cuisines. Asian food is the most well-liked subcategory of this cuisine (Ferguson 2003).

The cuisine can be classified as national, international, or traditional depending on location the cuisine comes from. One of the most complex culinary traditions is undoubtedly national cuisine. The question of what constitutes one's national cuisine is highly contentious. The issues arise because the territorial feature is not the only factor that defines the national cuisine. It is influenced by numerous other elements, including historical, political, religious, economic, and geographic aspects. These influences are demonstrating to us that the national cuisine is not a static category but rather is evolving as a result of them. It will take a long time for these changes to occur, and they are occurring over an extended period of time. Individual preferences are among the most significant impacts of society (Fratkin et al. 1999; Suhaimi and Zahari 2014). It indicates that every person has distinct tastes and interests that are unique to them. The majority of these preferences in a society are formed during the growing up and schooling process by parents, friends, family, and other individuals who mold an individual's taste to suit the tastes of the community. More people may experience this change in flavor at the same time, which could further impact the community's shift in taste. This results in a change in the national cuisine. One crucial point to remember when discussing a national cuisine is that its primary characteristics now may not necessarily translate to those in ten, fifty, or even a hundred years (Cwiertka 2005).

Because food and eating customs are ingrained in people's cultures, they are incorporated into the cultural values that tradition encompasses. Furthermore, one might discuss traditional food ways, which include the types of food consumed, the habits of food consumption, and the methods and approaches to food preparation and cooking that have been handed down from one generation to the next. In terms of being maintained via interaction and education, traditional cuisine is comparable to national cuisine. Parents make the decisions about what and how much food a child should eat. A person's habits and food preferences have already been shaped by the impact of his family and his surroundings (school, faculty, workplace, food at events, etc.) before he is able to make decisions later in life. It is unlikely that the food ways can be changed at this time (Ferguson 2010). Traditional cuisine always refers to the customs that have been passed down from one generation to another, but national cuisine refers to the present foodways, including recently recognized ones. This is what distinguishes traditional cuisine from national cuisine. The new lifestyles brought about by advancement and modernity are the cause of these changes. The globalization process, which brings civilizations closer together, is a prime example. This process led to the introduction of new goods into various cuisines, which gradually influenced their cooking practices and altered traditional foodways, eventually becoming a part of a person's national cuisine (Kocevski and Risteski 2020).

If we discuss about cuisine based on types of diets than these culinary classifications result from the various diets that people follow. They make reference to certain guidelines about what can be eaten, how it should be consumed, cooking and preparation methods, consumption times, product exclusions, and other topics. Due to shifting views on what constitutes health and how food affects our bodies, these kinds of cuisines are continuously expanding. Vegetarian, macrobiotic, diabetic, and other health-related cuisines, as well as the presently fashionable chrono and keto cuisines, are a few examples of diet-based cuisines. Some of these diets, like vegetarianism and diabetic cuisine, are consistent and unrelated to prevailing trends. These culinary classifications result from the various diets that people follow (Aydin and Corbaci 2019). They make reference to certain guidelines about what can be eaten, how it should be consumed, cooking and preparation methods, consumption times, product exclusions, and numerous other topics. Due to shifting views on what constitutes health and how food affects our bodies, these kinds of cuisines are continuously expanding. Vegetarian, macrobiotic, diabetic, and other health-related cuisines, as well as the presently fashionable chrono and keto cuisines, are a few examples of diet-based cuisines. Some of these diets, like vegetarianism and diabetic cuisine, are consistent and unrelated to prevailing trends. It's important to remember that while veggies can supply a lot of nutrients, many vitamins, including A, D, and K, as well as certain proteins and vital amino acids, are only present in animal products. Healthy food, healthy cuisine, and health-related cuisine are not the same thing. The primary reason is because the latter make reference to goods that are generally seen as healthy, yet some may be detrimental to a population that is afflicted with health problems. Fruit, for instance, is typically regarded as a healthy food, but if not taken as directed by diabetics, it might be fatal. Health-related cuisine refers to stringent dietary guidelines designed to keep an individual's present state of health from getting worse. A qualified physician or dietitian has recommended this diet (Van den Berghe 1984; Gheorghe and Bulin 2014).

## 7.3 Cultural Essence of Indian Cuisine

Indian customs and cuisines have been significantly impacted by the various cultures and religions who consider India home. Although vegetarianism is widely accepted in India, some areas—particularly those around the coast—are renowned for their non-vegetarian dishes. In India, rice and wheat are staple foods. Nonetheless, distinct cuisines are recognized in several Indian areas, including Gujarat, West Bengal, Maharashtra, Punjab and Southern India. An essential component of Indian cooking is bread (Srinivas 2011). The three most common types of bread are naan, paratha, and roti (flat bread). A specialized oven known as a tandoor oven is used to cook naan. In India, parathas can be eaten plain or packed with potatoes. In different parts of India, certain drinks have distinct names. In India, tea and coffee are widely consumed. Cardamom, ginger, clove, and other spices are being prepared with tea leaves to make a unique drink known as masala tea. India is renowned for its herbs and spices (Mangalassary 2015). When Christopher Columbus attempted to travel to India in search of spices, he discovered the new world. The blend of spices and herbs gives Indian traditional meals their robust flavor. Snacks are typically served with tea or coffee and tamarind or mint sauce in Indian cuisines. Condensed milk, grain flours, lentils, chickpea flour, semolina, and vegetables like pumpkins and carrots are among the ingredients. Various kinds of confections, including mithai, halwa, kheer, payasam, and sevai are made and served during religious holidays (Antani and Mahapatra 2022).

Indian cuisine also uses a range of legumes, including lentils and dried peas and beans. The soup-like consistency of daals is achieved by using lentils. Red lentils, black beans, chickpeas, pigeon peas and green grams are a few common lentils and legumes. These can be turned into flour or used whole or divided in curries. Vegetables are coated with lentil flour prior to frying. A popular ingredient in batter mixes for deep-fried foods is chickpea flour. In their culinary traditions, vegetables are an essential component. They can be included in a soup, snack, appetizer, main course, or side dish. They are also utilized in the production of various pickles, or achaar. Okra, eggplants, green peas, and potatoes are a few common veggies. Every meal can include vegetables (Shaikh et al. 2024). To transform them into curries, they are usually fried with oil and various spices. Indian also consume a variety of fruits as snacks or during meals. Mangoes are often consumed, either raw or pickled. They are also utilized in a number of cocktails and sweets. Using a range of herbs and spices is a hallmark of Indian cooking. Asafoetida, black mustard seed, cumin, coriander, turmeric, fenugreek, ginger, garlic and chilli pepper are among the often used spices. Garam masala, a powder that usually contains dried spices like cardamom, cinnamon, and cloves, is a common spice blend. This blend of spices differs depending on the location (Rathore and Shekhawat 2008). In many aspects, South Asians' eating habits are comparable. Three meals plus an evening snack often make up the meal plan. Families congregate in the evenings for tea and a snack. The main course is consumed well into the evening.

## 7.4 Cultural Essence of Arab Cuisine

Arab customs and culture are embodied in their traditional cuisine. The cuisine is a representation of hospitality, particularly traditional cuisine. The cooking techniques employed in traditional Arab cuisine differ depending on the specific type of food. When choosing traditional dishes, consideration should also be given to the food's ingredients and nutritional value. Stir-frying is the first cooking technique, and then boiling or simmering comes next. Arab traditional cooking techniques also include roasting or baking in a Tannor, a type of traditional oven. A portion of the traditional Arab cuisine is influenced by Mediterranean cuisine, which is rich in nuts, grains, fruits, vegetables, fish, and olive oil. The Maghreb is a unique geographical location in northwest Africa that is part of the Arab culture (Rosenberger 1999). It encompasses Tunisia, Mauritania, Algeria, Morocco, and Libya and is sometimes referred to as the Berber world. Food in the Maghreb is prepared using a variety of techniques, such as pickling, drying, salting, and fermentation. The Maghren people use olive oil as a medication and in their food, just like the Mediterranean people do. Roasted olives are used to make the unique oil known as alouana (Nabhan 2008). Argan oil is also widely used in the Maghreb region. The predominant diet in the Maghreb is couscous. Rice is replaced in the diets of other Arab nations by this carbohydrate-based cuisine. Couscous can be either vegetarian or eaten with beef or chicken. Pasta, chickpeas, barley, and other ingredients are used to make other typical Maghreb recipes. Dairy products with fermentation occupy a special place on the list of traditional Maghreb foods. Examples include fermented camel milk and its byproducts. Northwest Africa is the origin of the tagine culinary preparation. The word "Tagine" has the sound of the Maghreb when it is read or heard. A combination of meat, poultry, or fish, vegetables, and spices may be included in tagine, which is prepared using specialized utensils. Within the Maghreb countries, there are minor variations in tagine preparation. A variety of sweets and desserts, including samct, raisins, pastilla, baklawa, and other traditional foods, are used to flavor Maghreb cuisine (Gvion 2006; Al-Khusaibi 2019).

## 7.5 Cultural Essence of African Cuisine

In terms of both geographical area and population, Africa is the second-largest continent after Asia. Although it is primarily tropical, there are savannas in some areas. North Africa, which is comprised of the Maghreb nations; Sahel, which is composed of Chad, Mali, Mauritania, Niger, and Sudan; West Africa, which comprises Benin, Burkina Faso, Cameroon, Cape Verde, Cote d'Ivoire, Gambia, Guinea, Guinea-Bissau, Liberia, Nigeria, Senegal, Sierra Leone, and Togo; and Central Africa, which comprises Equatorial Guinea, Gabon, São Tomé and Príncipe, and South Sudan (Cusack 2000). There is a dearth of written material on the subject, and African cuisine is not well-known abroad. In Africa, the main sources of raw materials are cereals, legumes, and tuber roots. More than 2000 types of grains, legumes, fruits, vegetables, roots, and tubers, as well as oil crops, are found across the continent. In most of Africa, leafy green vegetables also play a significant role in meals. Furthermore, fermented foods and drinks, whether alcoholic or non-alcoholic, are a significant component of African cuisine. These foods are manufactured both on a modest industrial scale and domestically (Lyons 2007).

Grains make up a large portion of Northern Africa's diet and are used to make porridge and flat breads. The main course at lunch, which is the main meal, is frequently couscous, which is prepared from hard wheat and millet. Important staples include legumes including black-eyed peas, yellow peas, lentils, and broad beans. It's popular to cook using garlic, onions, and olive oil. Oranges, lemons, pears, and mandrakes are among the fruits eaten and notable spices include cumin, caraway, clove, and cinnamon. Since the food is primarily influenced by the Islamic customs of the area, alcoholic beverages are prohibited in many places (Okech and Timothy 2023). The main course in Central and Eastern Africa is made with the plentiful plantains and green bananas. This can be consumed with a delicacy made from cowpea, cassava, or pumpkin leaves and enhanced with red palm oil or groundnut sauce. Rice, potatoes, and a thick porridge made from cooked maize are additional mainstays. In Southern Africa, a staple crop—typically rice or maize—served with stew is the focal point of the traditional meal. Depending on the nation, the most popular cornmeal dish is known as mealie meal, pap, or nshima. It is typically consumed with stew poured over it, which may contain turnips, carrots, spinach, or cabbage that has been boiled. The southern portion of the continent offers a wide range of fruits. Due to its cultural diversity, South Africa is of particular relevance within this region. The diets of South Africans of Dutch, other European, and Asian Indian ancestry are comparable to those of their home nations (Holloway 2006).

## 7.6 Cultural Essence of Chinese Cuisine

China is divided into several areas, including Inner Mongolia, Guizhou, Hunan, Yunnan, Xinjiang, Gansu, Ningxia and Chengdu. Every region has a distinctive traditional method of preparing food. Rice is still a crucial basic food (Erlich 2004). There are several ways to eat rice, including cereals, noodles, puffed rice, cakes, fermented rice, and snacks made from extruded rice. The second significant commodity is seafood, particularly jellyfish, which plays a significant role in Chinese customs and is a staple dish during social gatherings. Chinese vinegar and fermented dairy products, such as tofu, are well-liked traditional meals. Chinese traditional cuisine also frequently includes a variety of soups (Li and Hsieh 2004; Nishizawa 2022).

## 7.7 Cultural Essence of Australian Cuisine

Native Australians have never practiced intensive agriculture; instead, they transitioned from hunting and gathering to an industrial system of farming, processing, and transportation using machinery. According to researchers, practicing intensive agriculture is essential to producing high-quality cuisine. For this reason, plant-based diets are regarded as supplements to the majority of meat-based diets (Bannerman 2011). Bush food is the term for the plant-based diet consumed by Aboriginal Australians. These foods are still widely consumed in Australia, and people there firmly feel that bush food knowledge is valuable and should be preserved. The Aboriginal people used about 5000 different types of traditional foods. The Mediterranean peasants who migrated to Australia following World War II also had an impact on Australian cuisine (Santich 2006). The two most popular commercial bush foods are akatyerr and certain acacias. Spices include things like Wattleseed, Anise Myrtle, Lemon Myrtle, and Tasmannia pepper (Petruso 2024). Bush tomatoes (*Solanurn chippendalei*), raisins, quandong (*Santalum acuminatum*), native figs (*Ficus platypoda*), wild oranges (*Capparis* spp.), desert bananas (*Leichhardtia australis*), and kakadu plums are a few examples of indigenous fruits and vegetables. Yams (*Dioscorea* species), wild potatoes (*Ipornoea costata*), native onions (*Cyperus* species), and yam daisies (*Microseris scapigeris*) are among the root and tubers. One of the most popular drinks among the native Australians is coconut milk. Additionally, plant segments such as flowers are steeped in water for making drinks. Corn, potatoes, beans, and squash are staples of traditional Native American diets (Luckins 2013).

## 7.8 Traditional Plant-Based Diets and Their Contributions to Health and Sustainability

Traditional plant based meals are high in bioactive compounds since most of them are prepared with herbs and spices. A variety of phytoconstituents, including phenolic acids, flavonoids, tannins and quinones, as well as bioactive peptides, protein, vitamins, minerals, organic acids, and bacteriocins are found in traditional diets. Traditional fermented foods, primarily soy products, exhibit potent antioxidant action that chelates metal ions and prevents lipid peroxidation. For instance, fermented soy products like tofu and sufu exhibit strong antioxidant qualities. Fermentation enhances reducing power, radical scavenging activity, and overall antioxidant activity (Xiao et al. 2015). Additionally, fermentation aids in the synthesis of the short chain, low molecular weight components that give rise to bioactive peptides. This is due to the fact that the molecular weight distribution and type of amino acids in the chain are related to the antioxidant property. However, because of their place of origin, raw materials, processing technique, and storage conditions, regional foods' nutritional worth and functioning vary greatly (Wang et al. 2003).

Red lentil, Chickpea, Black gram, Red gram and Green gram are the most widely consumed pulses in South Asian countries. India, Bangladesh, China, Russia, Canada, the United States, Brazil and Australia account for the majority of global pulse production. Protein, vitamins, minerals, and polyphenols can all be found in good amounts in pulses. They are used to make bread, snacks, purees, soups, chutneys, and sweets (Sozer et al. 2017). They can also be roasted, fried, or boiled with spices. The two most popular pulse processing techniques that have historically increased iron absorption are germination and malting. This is because it aids in increasing the amount of Vitamin C or lowering the levels of phytic acid and tannin (Hemalatha et al. 2007). Furthermore, because sprouting reduces the tannin concentration, it increases the bio accessibility of iron. Legumes' fermentation also aids in lowering their oligosaccharide fractions, which helps with gas, diarrhea, and digestive issues. Microorganisms break down these oligosaccharide fractions into useful mono- and disaccharides during fermentation. For instance, during tempe (beans) fermentation, the trypsin inhibitor is inactivated and the proportions of multiple oligosaccharides are decreased. According to reports, people who regularly consume Japanese traditional miso soup (soy bean barley paste) may be at lower risk of developing stomach cancer (Hirayama 1982).

Rice (primarily parboiled rice, flaked rice, hand-pounded rice, puffed rice, and rice wine), whole wheat flour (primarily chapati, puri, bread, naan, biscuit, idli, dosa, noodles, vermicelli, and sweet food), maize (primarily ogi, kenkey, and mawe), finger millet, and sorghum are the main ingredients of cereal-based traditional meals. The primary source of calorific value, which is primarily derived from protein and starch, is cereals (Tontisirin et al. 2002).

For instance, the fermented rice and black gram product known as idli has a wide variety of nutritional benefits. *Streptococcus faecalis, Lactobacillus mesenteroides, Lactobacillus plantarum* and *Lactobacillus lactis* are the main microorganisms that contribute to the leavening process of idli. These microbes synthesize vitamin B12 and the beta-galactosidase enzyme, which enhance the probiotic's qualities (Iyer et al. 2013). Additionally, two traditional Japanese rice products—vinegars and komesu—have been shown to have the ability to inhibit lipid peroxidation, prevent hypertension, lower blood pressure, and slow the growth of tumors. Additionally, another traditional Indian dish made from fermented millet, ambali, is high in minerals and thiamin, tryptophan, and riboflavin. Organic acids (primarily butyric, heptanoic, formic, propionic, lactic, caprylic, palmitic, and myristic), alcohols (primarily ethanol, amyl alcohol, and isoamyl alcohol), aldehydes and ketones (primarily formaldehyde and acetaldehyde), and carbonyl compounds (primarily furfural, methional, and glyoxal) are the main compounds that are produced during the fermentation of cereals. Moreover, the flavor of fermented food is derived from a small number of volatile chemicals that are generated during grain fermentations. The presence of acetic, butyric, and diacetyl acids gives fermented cereal-based products a distinctive odor. Nonetheless, trypsin inhibitors may be present in some foods made from cereals, but they can be eliminated using conventional processing techniques, primarily soaking, heating, and fermentation (Murooka and Yamshita 2008). Flavonoids, carotenoids, vitamins, minerals, and other phytochemicals found in abundance in fruits and vegetables are beneficial to human health. Anthocyanin, tannins, organic acids (malic acid, citric acid, oxalic acid, tartaric acid, and fumaric acid), geraniol, eucalyptol, essential oils, vitamins, and minerals (K, Cu, Zn, and Mg) are all present in mulberry fruit juice, for instance, and may help prevent coronary heart disease (Gupta and Meghwal 2021). Additionally, in Korea, Japan, and China, persimmon and its leaves are utilized as traditional tea and fruit, respectively. This is due to the fact that persimmon and its leaves are employed in traditional medicine and include flavonoids, tannins, vitamins, and minerals. In addition, jackfruit is a traditional fruit that is widely eaten in Asian nations and is rich in vitamins, minerals, carotenoids, phenolic compounds, and antioxidants (Kennedy et al. 2021). Conversely, the commonly available vegetables worldwide is the cucumber, which is rich in lignans, glucosides, apigenin, and cucurbitacins. One of the most popular techniques that has long been used to preserve cucumbers is fermentation. Similar to this, kimchi, or fermented cabbage, is another fermented vegetable that is popular in Korea because of its probiotic, anti-oxidative, anti-aging, anti-cancer, and anti-constipation qualities. Minerals (Ca, Fe, and K), vitamins (ascorbic acid, B-complex, and carotene), dietary fiber and amino acids (lysine, arginine, glutamic acid, aspartic acid, and alanine) are present in kimchi (Kristbergsson and Ötles 2016).

## 7.9 Conclusion

Cuisine is a manifestation of culture, identity, tradition, and way of life. To maintain and increase market share, cultural foods' quality level—that is, their safety, processing and preparation, and health—is essential. Classic meals are made using ingredients, cooking techniques, and eating habits that have been transferred through the centuries and are associated with a particular region. These show how different countries' cultures, beliefs, and religions, as well as their food preferences, customs, and resource availability, vary. The cultural cuisines of various nations varies based on geographic location and ethnicity. These become more valuable when they cross borders because they are more distinctive. Thus, it is crucial to first record the features of these foods, including their components, methods of processing or preparation, unique qualities, and preservation.

## References

Albala K (2000) Southern Europe. Cambridge World Hist Food 2:1203–1210

Al-Khusaibi M (2019) Arab traditional foods: preparation, processing and nutrition. In: Traditional foods: history, preparation, processing and safety. Springer International Publishing, Cham, pp 9–35

Anderson EN (2014) Everyone eats: understanding food and culture. In: Everyone eats. New York University Press

Antani V, Mahapatra S (2022) Evolution of Indian cuisine: a socio-historical review. J Ethnic Foods 9(1):15

Aydin B, Corbaci A (2019) The cuisine and its historical development. Co-Editors, p 153

Bannerman C (2011) Making Australian food history. Aust Humanit Rev 51:49–63

Counihan C, Van Esterik P (2013) Food and culture, vol 367. Routledge, New York

Cusack I (2000) African cuisines: recipes for nationbuilding? J Afr Cult Stud 13(2):207–225

Cwiertka K (2005) Culinary culture and the making of a national cuisine. In: A Companion to the Anthropology of Japan, pp 415–428

Erlich R (2004) Cultural and historical trends and influences of food, nutrition and cuisine on health and development. Asia Pac J Clin Nutr 13(2)

Ferguson PP (2003) BELLY TALK: Gastronomie, gastrolâtrie, and gourmandise in the 19th century. Dix-neuf 1(1):1–15

Ferguson PP (2010) Culinary nationalism. Gastronomica 10(1):102–109

Fratkin EM, Roth EA, Nathan MA (1999) When nomads settle: the effects of commoditization, nutritional change, and formal education on Ariaal and Rendille pastoralists. Curr Anthropol 40(5):729–735

Gheorghe G, Bulin D (2014) Cuisine-a regional tourism differentiation tool. Knowl Horiz Econ 6(2):194

Gupta R, Meghwal M (2021) Fermented food based on cereal and pulses. In: Advances in cereals processing technologies. CRC Press, pp 153–217

Gvion L (2006) Cuisines of poverty as means of empowerment: Arab food in Israel. Agric Hum Values 23(3):299–312

Hemalatha S, Platel K, Srinivasan K (2007) Influence of germination and fermentation on bioaccessibility of zinc and iron from food grains. Eur J Clin Nutr 61(3):342–348

Hirayama T (1982) Relationship of soybean paste soup intake to gastric cancer risk. Nutr Cancer 3(4):223–233

Holloway JE (2006) African crops and slave cuisine. Slave Rebellion Web Site. Available at: ricediversity.org/outreach/educatorscorner/.../African-Crops-and-Slave-Cuisine. doc (accessed 31 Oct 2025)

Iyer BK, Singhal RS, Ananthanarayan L (2013) Characterization and in vitro probiotic evaluation of lactic acid bacteria isolated from idli batter. J Food Sci Technol 50(6):1114–1121

Kennedy G, Kanter R, Chotiboriboon S, Covic N, Delormier T, Longvah T, Maundu P, Omidvar N, Vish P, Kuhnlein H (2021) Traditional and indigenous fruits and vegetables for food system transformation. Curr Dev Nutr 5(8):nzab092

Kocevski J, Risteski M (2020) Defining cuisine and types of cuisine. Horiz Int Sci J Ser A Soc Sci Humanit 26:291–302

Kristbergsson K, Ötles S (eds) (2016) Functional properties of traditional foods, vol 12. Springer, New York, p 396

Li JR, Hsieh YHP (2004) Traditional Chinese food technology and cuisine. Asia Pac J Clin Nutr 13(2):125–130

Luckins T (2013) Historiographic foodways: a survey of food and drink histories in Australia. Hist Compass 11(8):551–560

Lyons D (2007) Integrating African cuisines: rural cuisine and identity in Tigray, highland Ethiopia. J Soc Archaeol 7(3):346–371

Mangalassary S (2015) Indian cuisine—the cultural connection. In: Indigenous culture, education and globalization: critical perspectives from Asia. Springer Berlin Heidelberg, Berlin, Heidelberg, pp 119–134

Messer E (1984) Anthropological perspectives on diet. In: Annual review of anthropology, pp 205–249

Milton K (2000) Hunter-gatherer diets—a different perspective12. Am J Clin Nutr 71(3):665–667

Murooka Y, Yamshita M (2008) Traditional healthful fermented products of Japan. J Ind Microbiol Biotechnol 35(8):791

Nabhan GP (2008) Arab/American: landscape, culture, and cuisine in two great deserts. University of Arizona Press

Nishizawa H (2022) On the formation of chinese national cuisine: historical and anthropological perspectives. In: Making food in local and global contexts: anthropological perspectives. Singapore, Springer Nature Singapore, pp 187–208

Okech RN, Timothy DJ (2023) Culinary traditions and heritage foods in Africa. In: Cultural heritage and tourism in Africa. Routledge, pp 154–169

Petruso M (2024) Wild food: colonialism and the construction of Australian cuisine. Cookbooks, cultivation and the taste of home. Cult Hist: Student Res Papers 8(2):13–32

Rathore MS, Shekhawat NS (2008) Incredible spices of India: from traditions to cuisine. Am-Eurasian J Bot 1(3):85–89

Renna M, Rinaldi VA, Gonnella M (2015) The Mediterranean diet between traditional foods and human health: the culinary example of Puglia (Southern Italy). Int J Gastron Food Sci 2(2):63–71

Rosenberger B (1999) Arab cuisine and its contribution to European culture. In: Food: a culinary history, pp 207–223

Santich B (2006) The high and the low: Australian cuisine in the late nineteenth and early twentieth centuries. J Aust Stud 30(87):37–49

Shaikh M, Sunooj KV, Rahman MH, Navaf M, Ali TM (2024) Lentils: a recent review on global trade and popular regional cuisines. Legume Sci 6(3):e252

Smith K (2005) From milk to maize: the transition to agriculture for Rendille and Ariaal pastoralists. In: As pastoralists settle: social, health, and economic consequences of pastoral sedentarization in Marsabit District, Kenya. Springer US, Boston, MA, pp 137–153

Smith R, Smith PA (2003) An assessment of the composition and nutrient content of an Australian Aboriginal hunter-gatherer diet. Aust Aborig Stud 2:39–52

Sozer N, Holopainen-Mantila U, Poutanen K (2017) Traditional and new food uses of pulses. Cereal Chem 94(1):66–73

Srinivas T (2011) Exploring Indian culture through food. Educ About Asia 16(3):38–41

Suhaimi MZ, Zahari MSM (2014) Common acceptable cuisine in multicultural countries: towards building the national food identity. World Acad Sci Eng Technol, Int J Soc Behav Educ Econ Bus Ind Eng 8(3):859–865

Tontisirin K, Nantel G, Bhattacharjee L (2002) Food-based strategies to meet the challenges of micronutrient malnutrition in the developing world. Proc Nutr Soc 61(2):243–250

Van den Berghe PL (1984) Ethnic cuisine: culture in nature. Ethn Racial Stud 7(3):387–397

Wahlqvist ML (2007) Regional food culture and development. Asia Pac J Clin Nutr 16:2

Wahlqvist ML, Kouris-Blazos A, Hsa-Hage BH (1997) Aging, food, culture and health. Southeast Asian J Trop Med Public Health 28:100–112

Wang L, Saito M, Tatsumi E, Li L (2003) Antioxidative and angiotensin I-converting enzyme inhibitory activities of sufu (fermented tofu) extracts. Japan Agric Res Q 37(2):129–132

Xiao Y, Xing G, Rui X, Li W, Chen X, Jiang M, Dong M (2015) Effect of solid-state fermentation with Cordyceps militaris SN-18 on physicochemical and functional properties of chickpea (Cicer arietinum L.) flour. LWT-Food Sci Technol 63(2):1317–1324

# Chapter 8
# Education and Advocacy for Plant-Based Eating

**Abstract** Significant health benefits can be obtained from a plant-based diet that emphasizes plant foods (vegetables, fruit, pulses, grains and nuts) and limits consumption of refined processed foods and animal products (meat, poultry, fish, dairy and eggs). Obesity, diabetes, heart disease, and some types of cancer can be prevented by eating a plant-based diet. Eating behaviors and food habits are acquired during childhood and adolescence and typically carry over into adulthood. Therefore, eating a plant-based diet can help them make healthier food choices later in life in addition to being a significant dietary quality indicator. Research on the prerequisites and procedures involved in changing diets to a more plant-based diet is lacking, despite the potential advantages of a plant-based diet. Hence, the present chapter highlights the significance of education, advocacy and current frameworks of plant based eating to achieve universal food security and to create a habitable planet for all living things.

**Keywords** Advocacy · Current Frameworks · Eating Habits · Education · Plant Based Diet

## 8.1 Introduction

Significant environmental effects (such as land usage, biodiversity loss, and climate change) and an inadequate response to human nutritional demands are associated with the existing global food system. The food system has to change to more sustainable and easily accessible dietary patterns that support human health and the planet's ecological security, according to recent research and publications. Plant-based foods often have a lesser environmental impact than those derived from animals. This includes less usage of freshwater and land, lower greenhouse gas emissions, and a lower risk for eutrophication and acidification. Therefore, a shift toward more plant-based diets, particularly in more economically developed nations, supports the transformation of the food system by relying on lower consumption of animal products and refined carbohydrates and higher consumption of lower-impact

 
N. Patel et al., *The Future of Food: Exploring Edible Plants for a Sustainable Diet*,
https://doi.org/10.1007/978-3-032-15027-1_8

foods like fruits, vegetables, and whole grains (Bhat et al. 2019). Through the involvement of children and adolescents from diverse cultural and socioeconomic backgrounds, as well as their families and communities, school meals have been suggested as a pertinent platform to help speed sustainable food transitions. In the global food chain, the school meal system is important since it provides daily meals to some 370 million children worldwide (Shepon et al. 2018). Additionally, school meals have a multifaceted impact on society, ranging from immediate advantages for education and public health to social protection, agricultural and economic output, and environmental sustainability. These improvements won't happen on their own, even while the school meal program has the potential to facilitate food sustainability transitions (Springmann and Freund 2022). For instance, it has been shown that significant obstacles to change, particularly in the school setting, include inadequate plant-based meal offerings, meat-centric portrayals of a proper meal, and a lack of food knowledge. To lessen the excessive production and consumption of products derived from animals, a number of meat curtailment measures have been proposed, including institutional commitments, taxes and subsidies, and public laws (Stoll-Kleemann and Schmidt 2017).

Coordination between governing bodies, market participants, and civil society is necessary for the development and implementation of those policies. Support from the organizations and persons who will ultimately be impacted by the changes, whether directly or indirectly, is also necessary (Kos and Jerman 2019; Roque et al. 2022). Recent studies, however, indicate that attempts to advance sustainable food transitions at the practice and policy level may be jeopardized by political and ideological polarization, and that both governmental and non-governmental organizations may need to validate the significance of meat reduction in various meal settings, such as canteens and restaurants. To our information, no studies have yet to systematically examine teachers' and families' perceptions of school-based policies aimed at reducing meat consumption and encouraging a greater plant-based diet, despite the fact that they have been identified as stakeholders in the midday meals system. The dearth of data on this topic makes it more difficult to match school community expectations and requirements with sustainable food changes, which could potentially cause those transitions to be delayed because of low engagement (Graça et al. 2022; Meier et al. 2022).

## 8.2 Why Plant Based Diet Matter for Education?

According to conservative projections, humans consume over 65 billion land-based animals annually, and fishing and aquaculture that include wild capture result in the death of almost a trillion (and possibly more) fish annually. Animal consumption on this scale is unmatched in human history. However, because these activities are conducted out of the public eye, usually in massive trawler nets in the middle of the ocean or windowless concentrated animal feeding plants, most people are ignorant of the scope of animal farming, trawling, and slaughter, as well as the brutalities involved (Mood and Brooke 2012). Even in wealthy nations with the highest per unit animal consumption, there is little media coverage of the issue, and national governments also say nothing about the suffering of animals raised for food. In actuality, lobbying for meat, dairy, and animal-based foods have a lot of political clout, and laws may be in place in some countries that make it unlawful to conduct undercover investigations or whistleblowing in these areas. These factors work together to reduce public awareness of the size of these sectors and the cruel circumstances used to grow, slaughter, and process animals. Several studies have found that the best way to mitigate the suffering of farmed animals, as well as the extinction of wild animals, environmental harm to soil, water, and air, and global food insecurity, is to switch to a plant-based diet (Aaltola 2019).

Numerous legal scholars have discussed how slowly laws can affect social change and how unlikely it is that they will serve as the impetus for social change on a given topic. 8. While the legal system is essentially conservative, judges follow not only existing legislation and legal precedent but also prevailing cultural norms and social opinion that serve as the foundation for such legislation and precedent (Foer 2019). This is true even though law shapes culture and occasionally ushers in revolutionary social change by enacting new legal prohibitions or repealing old ones. Judges have been known to use shifting social views to support departing from established legal precedent when they forbid or approve novel ideas they know to be contentious. Furthermore, legislative change happens when political will arises, usually in response to public opinion. The courts and legislatures have so far proven to be unsuccessful avenues for significant change for the vast majority of animals, despite the fact that other social movements aimed at challenging structural injustices and ingrained cultural norms have had success in similar arenas (Alvaro 2020). The common law was founded as and continues to be a highly anthropocentric legal system. Redesigning current animal-based food systems has not received government or other mainstream policy attention despite increased coverage of animal agriculture's links to climate change and other harmful environmental phenomena or its increased potential to cause fatal zoonotic pandemics. Government responses to climate change have instead focused on greening transportation and buildings while notably ignoring the need to curtail animal husbandry (Kemmerer 2019). The majority of governments support these industries with subsidies and through agricultural ministries that regularly collaborate with agricultural stakeholders to thwart any attempts, even within their own government, to undermine the meat, dairy, and egg industries through the policy work of other ministries and departments (Crowe 2019).

Although the UN made a clear connection between animal agriculture and zoonotic diseases in late 2020, alternative media outlets, not national governments, have largely examined how animal agriculture raises the world's risk of zoonotic pandemics in order to increase public awareness of the origins of COVID-19. The shutdown of slaughterhouses and meatpacking industries was not prompted by growing evidence that these establishments were hubs for the spread of COVID-19 among human workers (Damerell et al. 2013). In the name of "food security," we instead witnessed industry bailouts and animal agriculture designated as "critical infrastructure," while slaughterhouses that did temporarily close due to rising infection rates were ordered to remain open despite the continued health and mortality risks to a highly racially diverse and precarious immigrant workforce. The issue of "wet markets," which are portrayed as a foreign phenomenon, is the principal focus of government and mainstream media criticism of food practices in relation to the COVID-19 outbreak in both Canada and the United States (Whitfort 2021).

Without significant conceptual change and the incorporation of alternative valuations, it is difficult to see how common law or international law, with its unavoidable anthropocentric and colonial imprints, can address today's pandemic crises, climate disasters, and other global emergencies. Colonial human exceptionalism, which is based on private property logics and views the majority of people on the planet—both nonhuman and human—as naturalized and continual "resources" or "labor" to be exploited, is the primary cause of today's issues. In fact, it is doubtful that anything will motivate legislative discussion or policy action regarding the necessity of reorganizing food systems to move away from animal-based agriculture if the fears of irreversible climate change or another worldwide pandemic are insufficient (Aguirre and Orihuela 2013). An essential (although still lengthy) route to achieving change is educational intervention, rather than relying solely on the legal system to implement the required plant-based remedies. It would be incorrect to imply that adults are incapable of forming new habits or ways of thinking, or that behavioral changes in adults have not aided in significant social changes in the defense of human rights or other changes in standards and values. However, a wealth of research shows that although we may alter our behavior as adults, doing so is difficult and that most people who try to break long-standing habits—regardless of the particular behavior—fail. Changing one's diet is not an exception. Numerous behavioral and cognitive barriers have been identified by social psychologists and sociologists as preventing people from switching to a plant-based diet. Studies show that women are disproportionately represented among vegetarians and vegans and are also more inclined to contemplate switching to a plant-based diet, demonstrating the pervasiveness of gendered and heteronormative connections surrounding eating animal meat in particular in mainstream Western cultures (Chiu and Lin 2009). The obstacles are so great that research shows that even when people want to and are able to switch to plant-based diets, they are likely to quickly return to their old eating habits. Among these considerations, the "meat paradox"—the circumstance in which people claim to love animals but yet buy products from the sector despite knowing about the conditions of intensive farming—is particularly noteworthy. In this case, blocking-out techniques, denial, or justification are used to deal with cognitive dissonance. This shouldn't come as a surprise given that the majority of us can look back on our own early years and experiences to realize that food is about more than just consuming calories and staying alive; it's also about nostalgia, family, culture, community, and a sense of belonging. Even for adults who are receptive to learning about the origins of food, attempting to change adult behavior through education usually yields little in the way of immediate or long-term social change because of "denialism" and the cognitive and emotional defenses adults use against the facts and the larger social and political context. For this reason, whether through official curriculum changes or more unofficial channels, focusing on children's education makes sense as a primary goal in public awareness efforts. Young children already associate animals with themselves (Dılmaç et al. 2007). It is true that there are many animals in their world, and there is some evidence to suggest that youngsters do not exhibit human exceptionalist thinking to the same extent as adults. But eventually, kids begin to absorb the human ideals of their parents and the organizations in their environment. In fact, schools are a component of a broader anthropocentric cultural infrastructure that teaches kids to accept the daily brutalities committed against animal bodies and the normalization of human instrumentalization of other life forms. Youngsters are capable of taking in knowledge that is appropriate for their age, thinking critically about it, and making moral decisions (Bryant and Dillard 2020). Even while we cannot anticipate revolutionary change, such early instruction can plant the seeds for introspection, discussion, critical thinking, and potential future action; for this reason, it is worthwhile. Existing channels like schools, books, apps, television, and other visual media, as well as discussions with parents, can all be used to convey this information. A greater number of homes and classrooms could be reached by expanding the several humane education program models currently provided by non-profit organizations. Furthermore, general education literature clearly demonstrates that the lessons children learn in elementary school can influence their perspectives as adults, particularly if such lessons are reinforced at home and maintained throughout life. Observing industry engagement in using educational channels to market their products and a positive perspective of animal agriculture to children provides additional evidence of the impact that such early education can have on children's perceptions of farmed animals (Cole and Stewart 2016; Linné and Pedersen 2014).

## 8.3 Advocacy Initiatives for Plant Based Eating

The shortcomings of contemporary industrial agriculture particularly industrial animal agriculture are coming to light increasingly frequently. The first step toward transformation is raising public awareness. Initiatives at all governmental levels to enhance school nutrition and provide plant-based lunch options are essential first steps in educating the youth. Businesses are beginning to make greater investments in processed foods made from plants. This is an obvious sign that corporations generally agree that people would become more interested in plant-based diets in a capitalist society (Viroli et al. 2023). Because highly processed meals have a higher profit margin than unprocessed plant-based foods, companies are concentrating on developing plant-based meat and dairy substitutes as well as other highly processed plant-based foods. Since nonprofits can oppose agricultural special interest groups, they are essential to the progress of plant-based diets. The diverse range of viewpoints that numerous nonprofit organizations contribute to public discourse encourages change in various spheres and in various ways. When it comes to lobbying government representatives to support their projects, nonprofit groups are particularly effective (Beverland 2014). People are ultimately the ones who bring about change through their own decisions and beliefs. Individuals select their cause and look for various methods to contribute to the development of the society in which they live.

The impacts of eating foods derived from animals are becoming more widely known. Education initiatives are necessary to alter the public perception of sustainable and healthful diets because animal-based foods continue to be a major part of the Standard American Diet. Campaigns that emphasize people's health and well-being over the more ethereal idea of climate change or the divisive subject of animal rights will be more successful (Herpich et al. 2022). The majority of individuals care about their health and well-being; prolonging one's own life is a significant goal, at least in Western cultures. Campaigns for education and public awareness must cover two different subjects. First, greater proof that a healthy diet does not require animal-based items is required. Second, there must be chances and resources for individuals to learn how to cook plant-based meals. As humans move away from animal-based diets, plant-based meat substitutes may be crucial. Long-term dietary changes, however, will be easier for those who know how to cook a range of plant-based foods. Businesses that have made investments in highly processed plant-based foods, plant-based meat substitutes, and dairy substitutes will concentrate their marketing efforts on these products, which somewhat compromises the health message of promoting plant-based diets. Despite its harmful impacts on the environment and human health, industrial animal husbandry may be so successful because of its excellent marketing and powerful lobby (Lea et al. 2006). The strategies of industrial agriculture could teach nonprofits and activists devoted to human and environmental health how to better communicate their message. Although numerous studies have refuted the claims that cow's milk is essential for bone health and that red meat protein is preferable, these ideas are nevertheless widely accepted. This is mostly because the dairy and meat industries have superb and sophisticated marketing (Comerford et al. 2021).

Expert marketing firms will be able to create equally successful plans to encourage plant-based diets and alter public perception. Apart from the little market for meat and dairy substitutes, there isn't a significant industry supporting a plant-based diet strategy that might support extensive advertising efforts. Many stakeholders can create and carry out campaigns. Nonprofits may launch campaigns, schools can teach it, and social media influencers are also very important. It will be necessary to create and apply holistic approaches that take into account the different factors that influence food choices (Brummans et al. 2016). Diets based on animals are still common today. To alter this narrative, deliberate efforts will be required. Animal-based diets are linked to strength and masculinity in the prevailing Western culture, whereas plant-based diets are linked to weakness, femininity, and deviation from the norm. Therefore, one tactic to alter the perception and narrative surrounding plant-based diets is to use "positive role models, particularly sports stars or those engaged in physically demanding roles." Perceptions can be altered by sharing the experiences of individuals like Nick Squires, a successful powerlifter and devoted vegan (Sabaté and Soret 2014). The author believes that it is impossible to overstate the importance of trailblazers and influencers who strive to alter the perception of what constitutes a healthy diet. People often look up to role models, and celebrities and athletes can significantly alter public attitudes. Promoting plant-forward diets initially could be beneficial in order to overcome the opposition in many facets of society and make it simpler for people to begin (Vinnari and Vinnari 2014). Most people who are accustomed to eating a lot of animal-based meals find the idea of switching to a plant-based diet where animal products are banned permanently intimidating. Since each plant-based meal is a positive step, progress should be appreciated.

Campaigns for marketing and education must consider cultural differences. Campaign-shared recipes must take into account the cultural background of their intended audiences. Culture is important, and there is a huge diversity of vegan dishes available, as the Food Empowerment Project demonstrates in their recipe collections for vegan Mexican, Filipino, Lao, and Chinese foods. The global food system is experiencing market failure as a result of the existing capitalist market system's disregard for environmental and health consequences. To transition to a more sustainable food system, such externalities must be internalized (Orset and Monnier 2020). Regulations and policies that promote the internalization of environmental and social externalities brought about by the industrial food system should be pursued by the California government, the US federal government, and international organizations. Initiatives could include health or environmental taxes, fines, and subsidies. Foods derived from animals, for instance, can be subject to different taxes than foods derived from plants. Although California sales tax does not now apply to food, models with comparable content have been discussed and put into practice. For instance, in an effort to deter people from drinking sugar-sweetened beverages, the city of Berkeley imposed a sales tax on them in 2014. California already imposes a tax on alcoholic beverages. One tactic used on the demand side of the equation would be to tax foods derived from animals. This would push consumers to modify their behaviour, which would alter the supply (Laestadius et al. 2016).

The existing pricing of groundwater use does not guarantee the long-term preservation of groundwater levels and, consequently, the sustainability of the environment and society. Reforming water rights is essential in order to start down the path to a more equitable and sustainable future. Raising the price of water to guarantee its sustainability may reduce the profitability of producing foods derived from animals and, depending on market conditions, will push companies to look for alternatives. Initiatives known as "Climate Smart Agriculture" encourage responsible land and water management, which contributes to the development of a more sustainable agricultural sector. They do not alter the food system's pricing structure, nevertheless, and are supported by the government (Fanzo 2023). The actual cost of production and consumption must be factored into food pricing in order to rely on the invisible hand of the market. To guarantee an adequate supply of wholesome food for everyone, the author believes it would be better to internalize all costs into food prices and provide government subsidies for low-income households. In addition to guaranteeing social sustainability through widely accessible food aid, this would establish suitable market incentives for producers and consumers to make decisions about production and consumption that are in line with environmental sustainability (Harris 2015).

The huge and influential industry of commercial animal agriculture uses its size and clout to influence public opinion through extensive marketing efforts and the political landscape through extensive lobbying. The diversity and complexity of viewpoints within the nonprofit and activist fields is one of their strong points. Because many interest groups can operate in silos or even against one another, this complexity can sometimes work against the shared objective of promoting a plant-based diet (Ingram and Zurek 2019). This is particularly evident in the various vegan communities, where self-declared ethical vegans maintain that the only legitimate justification for going vegan is animal rights concerns and that any lifestyle that is not entirely vegan is intolerable. Many ethical vegans believe that health-based justifications for a plant-based diet are insufficient. As a result, many who are considering switching to a plant-based diet or making the first moves to cut back on their intake of animal products get alienated. If such lobbying efforts are successful and the number of people eating plant-based diets rises, then all causes will gain (James 2004). Regardless of whether people select plant-based diets due to concerns about animal rights or for their own health, animals are not killed. Regardless of the motivation for a plant-based diet, people's health improves with it. The human experience is what all activism and non profit organizations have in common (Schleicher and Töller 2024).

Recognizing the power of the subject's intersectionality can be beneficial since it can produce a thorough alignment in the struggle for change. The success of every endeavor will depend on showing compassion to activists who may have different priorities but ultimately have the same objective. Collaborations among many organizations, governments, businesses, and non profits will have a greater impact than any one project could. Even if the ultimate objective is still unattainable, perfectionism must be set aside and a more practical approach might be used to make faster progress (Johns and Eyzaguirre 2006; Hemler and Hu 2019).

## 8.4 Active Frameworks and Relevant Issues

The safety of plant based and traditional food items is a significant concern for food producers worldwide, even while the intellectual property associated with traditional food may be its distinctive identifying characteristic from a regulatory standpoint. Protection of consumers from consuming contaminated food is a significant food policy concern for nations, and the growth of the traditional foods industry depends on customer trust. Food contamination is becoming more likely due to globalization, which has also revealed weaknesses in national and international capacity to provide sufficient levels of food safety and quality for all foods, especially traditional foods (Echols 2001). Concerns about food safety are commonly used as grounds for laws and regulatory frameworks, trade restrictions and other policies that limit global trade. As a result, foods' increased safety can help them become more widely available in international marketplaces. The integrity and safety of lengthy, intricate supply chains that necessitate close coordination and collaboration on a global scale become even more crucial because foods' economic value tends to rise with distance from their place of origin (Williams et al. 2023).

The highest risk of foodborne disease is found in low- and middle-income nations, where unsafe water used for food preparation, unsanitary conditions during food production and storage, low literacy and educational levels, and a lack of food safety laws or their enforcement are all contributing factors (Trichopoulou et al. 2007). According to the WHO's "Estimates of the global burden of foodborne diseases," consuming tainted food causes up to 600 million illnesses annually—nearly one in ten people—and 420,000 fatalities. With 125,000 children dying each year from foodborne illnesses, children under the age of five are especially vulnerable. The biggest burden of foodborne illnesses is found in South-East Asia and Africa, as reported by WHO. Therefore, food standards must be established to provide sufficient protection for the consumers of the trading nations while avoiding trade barriers. The Codex Alimentarius Commission, an international intergovernmental organization, develops the minimum standards of food safety needed to carry out international trade (Grote and Kirchhoff 2001; Mansour 2008).

When thinking about how customary foods will be regulated in the future, it's critical to understand that, while customary goods and recipes have been regulated locally through social and cultural norms and practices that have evolved over time, as well as local laws and regulations pertaining to food safety, traditional foods must also abide by formal food safety regulations when they are transported outside of their place of origin (Gallagher and McKevitt 2019). However, due to their ongoing, demonstrated safe use, these legal conditions, when met, can result in a number of benefits and economic opportunities, as well as preferential treatment in some jurisdictions. Following several years of discussion and consultation, the European Parliament and Council passed the Novel Food Regulation EC 258/97, which applied to new foods and ingredients that had not, prior to May 15, 1997, been "previously used to a significant degree for human consumption" within the EU. These foods included foods made using innovative technology or foods sourced from novel sources, such as krill oil high in omega-3 fatty acids. Relatively few new food sources were initially authorized by the Regulation (Williams et al. 2023).

Even when exporting developing nations could show the stringent and costly regulatory process effectively prevented them from accessing the lucrative EU market for a variety of goods. The few successful applications from outside the EU that were approved under the Regulation are noni juice, chia seeds, and baobab. Within the European Union, the innovative foods law did not apply to Benecol, a margarine that included stanol esters and was the first "functional food" to be marketed in Europe in 1995. Other sterol meals, however, required regulatory approval because they lacked a track record of safe use (Bloom and Hinrichs 2011). The European Parliament rejected a 2008 recasting of the Novel Foods Regulation because of disagreements on the topic of animal cloning, and the debate went on. Finally, in January 2018, the new Novel Foods Regulation (EU 2015/2283) went into effect. A centralized evaluation and authorization process has been implemented, along with a more thorough definition of "novel foods" in the new regulation. Under the new system, conventional meals from third countries—those outside the EU—that come from primary production and have a track record of safe use may be considered novel foods for EU regulatory reasons. These consist of traditional cuisines prepared from algae, fungi, microbes, plants, and animals (such as insects). A food's history of safe use indicates that its safety has been established by compositional data and by the fact that it has been used consistently for at 25 years in the typical diet of a sizable portion of the people in at least half of the country (Gallagher and McKevitt 2019).

Thus, the Novel Foods Regulation may yet prove to be an example of a successful attempt to balance the two sometimes conflicting goals of ensuring that modern, science-based standards of food safety are met by recognizing certain traditional foods as "traditionally safe" and celebrating third country traditional food products (and the cultures from which they originate) by granting them access to the European market (Philipp and Zander 2025).

## 8.5 Conclusion

There is ample proof that plant-based diets can help both people and the environment in many ways. But meals derived from animals are still common. Foods derived from plants are generally kinder to both people and the environment than those derived from animals. They are a crucial component of the jigsaw, but they won't address every problem with food security. In order to achieve universal food security and to create a habitable planet for all living things and future generations, the promotion and education of plant-based foods should be implemented in tandem with all other significant projects.

## References

Aaltola E (2019) The meat paradox, omnivore's akrasia, and animal ethics. Animals 9(12):1125

Aguirre V, Orihuela A (2013) Short-term interventions that accomplish humane education goals: an international review of the research literature. In: Teaching compassion: humane education in early childhood. Springer, pp 23–31

Alvaro C (2020) Raw veganism: the philosophy of the human diet. Routledge

Beverland MB (2014) Sustainable eating: mainstreaming plant-based diets in developed economies. J Macromark 34(3):369–382. https://doi.org/10.1177/0276146714526410

Bhat R, Khajuria M, Mansotra DK (2019) A systematic review on global environmental risks associated with pesticide application in agriculture. In: Contaminants in agriculture and environment: health risks and remediation, vol 1, p 96

Bloom JD, Hinrichs CC (2011) Moving local food through conventional food system infrastructure: value chain framework comparisons and insights. Renewable Agric Food Syst 26(1):13–23

Brummans BH, Cheong PH, Hwang JM (2016) Faith-based nongovernmental environmental organizing in action: Veroes' campaigning for vegetarianism and mindful food consumption. Int J Commun (19328036) 10
Bryant C, Dillard C (2020) Educated choices program: an impact evaluation of a classroom intervention to reduce animal product consumption. OSF Preprints. https://doi.org/10.31219/osf.io/hecyk
Chiu TH, Lin CL (2009) Ethical management of food systems: plant based diet as a holistic approach. Asia Pac J Clin Nutr 18(4):647–653
Cole M, Stewart K (2016) Our children and other animals: the cultural construction of human-animal relations in childhood. Routledge
Comerford KB, Miller GD, Boileau AC, Masiello Schuette SN, Giddens JC, Brown KA (2021) Global review of dairy recommendations in food-based dietary guidelines. Front Nutr 8:671999
Crowe K (2019) The long road to a new Canada's food guide. CBC News 9
Damerell P, Howe C, Milner-Gulland EJ (2013) Child-orientated environmental education influences adult knowledge and household behaviour. Environ Res Lett 8(1):015016
Dılmaç B, Kulaksizoğlu A, Ekşlı H (2007) An examination of the humane values education program on a Group of Science High School Students. Educ Sci-Theory Pract 7(3)
Echols MA (2001) Food safety and the WTO: the interplay of culture. Science, and technology. Kluwer Law International, Dordrecht
Fanzo J (2023) Achieving food security through a food systems lens. In: Resilience and food security in a food systems context. Springer International Publishing, Cham, pp 31–52
Foer JS (2019) We are the weather: saving the planet begins at breakfast. Penguin UK
Gallagher J, McKevitt A (2019) Laws and regulations of traditional foods: past, present and future. In: Traditional foods: history, preparation, processing and safety. Springer International Publishing, Cham, pp 239–271
Graça J, Roque L, Guedes D, Campos L, Truninger M, Godinho C, Vinnari M (2022) Enabling sustainable food transitions in schools: a systemic approach. Br Food J 124(13):322–339
Grote U, Kirchhoff S (2001) Environmental and food safety standards in the context of trade liberalization: issues and options (No. 39). ZEF Discussion Papers on Development Policy
Harris SM (2015) The myth of sustainable food supply and the urgent need for radical change from competitive corporatism to sustainable stewardship. In: Strategies towards the new sustainability paradigm: managing the great transition to sustainable global democracy. Springer International Publishing, Cham, pp 141–154
Hemler EC, Hu FB (2019) Plant-based diets for personal, population, and planetary health. Adv Nutr 10:S275–S283
Herpich C, Müller-Werdan U, Norman K (2022) Role of plant-based diets in promoting health and longevity. Maturitas 165:47–51
Ingram J, Zurek M (2019) Food systems approaches for the future. In: Agriculture & food systems to 2050: global trends, challenges and opportunities. World Scientific, pp 547–567
James D (2004) Factors influencing food choices, dietary intake, and nutrition-related attitudes among African Americans: application of a culturally sensitive model. Ethn Health 9(4):349–367
Johns T, Eyzaguirre PB (2006) Linking biodiversity, diet and health in policy and practice. Proc Nutr Soc 65(2):182–189
Kemmerer L (2019) The interconnected nature of anymal and earth activism. Am Behav Sci 63(8):1061–1079
Kos M, Jerman J (2019) Gardening activities at school and their impact on children's knowledge and attitudes to the consumption of garden vegetables. Prob Educ 21st Century 77(2):270–291
Laestadius LI, Neff RA, Barry CL, Frattaroli S (2016) No meat, less meat, or better meat: understanding NGO messaging choices intended to alter meat consumption in light of climate change. Environ Commun 10(1):84–103
Lea EJ, Crawford D, Worsley A (2006) Public views of the benefits and barriers to the consumption of a plant-based diet. Eur J Clin Nutr 60(7):828–837

Linné T, Pedersen H (2014) Expanding my universe. Critical animal studies education as theory, politics, and practice. In: Sorenson J (ed) Critical animal studies: thinking the unthinkable, pp 268–282

Mansour M (2008) Codex and its competitors: the future of the global regulatory and trading regime for food and agricultural products. In: Hasler CM (ed) Regulation of functional foods and nutraceuticals: a global perspective. Blackwell Publishing, Ames, pp 377–389

Meier CL, Brady P, Askelson N, Ryan G, Delger P, Scheidel C (2022) What do parents think about school meals? An exploratory study of rural middle school parents' perceptions. J Sch Nurs 38(3):226–232

Mood A, Brooke P (2012) Estimating the number of farmed fish killed in global aquaculture each year. Fishcount org uk, p 1

Orset C, Monnier M (2020) How do lobbies and NGOs try to influence dietary behaviour? Rev Agric Food Environ Stud 101(1):47–66

Philipp SM, Zander K (2025) A conceptual framework and review explaining the role and potential of product marketing in preserving traditional cultural landscapes. Front Sustain Food Syst 9:1638485

Roque L, Graça J, Truninger M, Guedes D, Campos L, Vinnari M, Godinho C (2022) Plant-based school meals as levers of sustainable food transitions: a narrative review and conceptual framework. J Agric Food Res 10:100429

Sabaté J, Soret S (2014) Sustainability of plant-based diets: back to the future. Am J Clin Nutr 100:476S–482S

Schleicher K, Töller AE (2024) Dietary behavior as a target of environmental policy: which policy instruments are adequate to incentivize plant-based diets? Sustainability 16(6):2415

Shepon A, Eshel G, Noor E, Milo R (2018) The opportunity cost of animal based diets exceeds all food losses. Proc Natl Acad Sci 115(15):3804–3809

Springmann M, Freund F (2022) Options for reforming agricultural subsidies from health, climate, and economic perspectives. Nat Commun 13(1):82

Stoll-Kleemann S, Schmidt UJ (2017) Reducing meat consumption in developed and transition countries to counter climate change and biodiversity loss: a review of influence factors. Reg Environ Chang 17(5):1261–1277

Trichopoulou A, Soukara S, Vasilopoulou E (2007) Traditional foods: a science and society perspective. Trends Food Sci Technol 18(8):420–427

Vinnari M, Vinnari E (2014) A framework for sustainability transition: the case of plant-based diets. J Agric Environ Ethics 27(3):369–396

Viroli G, Kalmpourtzidou A, Cena H (2023) Exploring benefits and barriers of plant-based diets: health, environmental impact, food accessibility and acceptability. Nutrients 15(22):4723

Whitfort A (2021) COVID-19 and wildlife farming in China: legislating to protect wild animal health and welfare in the wake of a global pandemic. J Environ Law 33(1):57–84

Williams LB, Jones M, Wright PF (2023) Decolonising food regulatory frameworks: importance of recognising traditional culture when assessing dietary safety of traditional foods. Proc Nutr Soc 82(4):454–467

# Index





N. Patel et al., *The Future of Food: Exploring Edible Plants for a Sustainable Diet*,
https://doi.org/10.1007/978-3-032-15027-1

MIX
Papier aus verantwortungsvollen Quellen
Paper from responsible sources
FSC® C105338

If you have any concerns about our products,
you can contact us on
**ProductSafety@springernature.com**

In case Publisher is established outside the EU,
the EU authorized representative is:
**Springer Nature Customer Service Center GmbH**
**Europaplatz 3, 69115 Heidelberg, Germany**

Printed by Libri Plureos GmbH
in Hamburg, Germany